IMAGES
*of America*

# DRAPER

Lone Peak is the highest peak in the Wasatch Range and is located between Little Cottonwood Canyon to the north and Corner Canyon to the south. Other lower peaks are Enniss Peak and Bighorn Peak. Lone Peak has an elevation of 11,260 feet. Richard Bell Sr. of Riverton was the first person to successfully hike Lone Peak. Bell Canyon was named for him. (Courtesy of Draper Historical Society.)

ON THE COVER: This photograph from 1910 shows members of the Allen family with their faithful dogs at "sheep camp." The sheepherders stayed with the sheep when they went to graze on their summer range. They worked from sunup to sundown making sure that none went astray and guarded the animals from predators such as coyotes and mountain lions. (Courtesy of Draper Historical Society.)

IMAGES
*of America*

# DRAPER

Katherine L. Weinstein, PhD

ARCADIA
PUBLISHING

Published by Arcadia Publishing
Charleston, South Carolina

Printed in the United States of America

Library of Congress Control Number: 2022951329

For all general information, please contact Arcadia Publishing:
Telephone 843-853-2070
Fax 843-853-0044
E-mail sales@arcadiapublishing.com
For customer service and orders:
Toll-Free 1-888-313-2665

Visit us on the Internet at www.arcadiapublishing.com

*To the people of Draper*

# CONTENTS

# ACKNOWLEDGMENTS

Unless otherwise noted, all images appear courtesy of the Draper Historical Society. In fact, this book would not have been possible without the assistance and contributions of this organization. Members of the society gave me access to their collection of photographs and primary sources, assisted with research, and fact-checked my chapter drafts. Anne Covington came in on Saturdays to help me with research, and Draper Historical Society president Lynne Orgill assisted in tracking down photos. Draper Historical Society's three volume comprehensive history of the community, *People of Draper, 1849–1924, Sivogah to Draper City 1849–1977*, and *People of Draper 1849–1934*, provided the main source material for this book. Anne Covington, Esther Kinder, Linda Richins, and Paul Smith also shared their personal photographs with me.

In addition, special recognition must be given to the local people with ties to Draper who took the time to track down old photographs and shared their family stories for the book: Bill Allinson, Tonya Atkinson, Don Ballard, Doug Ballard, David Bean, Kathleen Ballard Day, LaRayne Day, Jean Hendricksen, Alona Holm, Lyn Kimball, Paula Summers, Ray and Helene Terry, Lorna Toone, and Holly Westbrook.

Members of Draper Visual Arts Foundation imparted valuable information and granted permission to include images of works of art in the book. The Church History Library of The Church of Jesus Christ of Latter-day Saints and the Utah State Historical Society also provided photographic images.

I wish to thank my editor Caroline Vickerson for her guidance and reassurance during the entire process of compiling images and writing the text. Special thanks is due to my friends and work colleagues who answered my questions about Church-related organizations and used their connections to the Draper community to locate more photos. My husband, David, helped immeasurably by scanning photographs and taking pictures of artifacts and historic buildings as well. I also thank my daughter for her patience during this project and my mother for inspiring my love of history in the first place.

# INTRODUCTION

The city of Draper, Utah, is situated in a cove in the southeastern corner of the Great Salt Lake Valley along the Wasatch Front. Tens of thousands of years ago, the waters of Lake Bonneville carved out the mountains and valleys of the area. The ancient lake left a mountain of sand on the southwestern boundary of the cove now known as Point of the Mountain. To the east of the cove, Lone Peak towers over the landscape, and the Traverse Mountains are on its southern edge. Fresh water from canyon streams combined with fertile valley soil made the area ideal for farming. The earliest human inhabitants of the Salt Lake Valley were the ancient people referred to today as the Fremont. In subsequent centuries, Native Americans such as the Utes, Paiutes, and Shoshone traversed the Draper area during their annual hunting and gathering treks. The name Sivogah, which means "willows" in Ute and Paiute, two related Native American languages of Utah, was given to the area.

Later, European explorers of the American West as well as mountain men and fur trappers passed through Sivogah but did not establish a permanent settlement. In 1850, settlers belonging to The Church of Jesus Christ of Latter-day Saints built the first cabin in the place they called South Willow Creek. The pioneer community was soon after named Draperville after William Draper Jr., the town's first church elder. As of the 1890 census, the name was shortened to Draper.

For the next 140 years or so, Draper could be characterized as a quiet farming community. Many of the townspeople worked small farms and kept a few head of cattle and sheep. Draper also supported various industries over the decades, including sand mining, canneries and later poultry farming on a large scale.

By the late 1930s, Draper was known as the "Egg Basket of Utah" and marketed eggs to consumers across the United States as well as troops in the South Pacific during World War II. The dominance of the egg industry in Draper began to ebb in the 1950s due to increased competition from egg producers in California. Since the early 1960s, the Intermountain Farmer's Association has operated a feed processing facility and store on the site of the Draper Egg Producers Association.

The 1950s and 1960s brought some unusual, if temporary, developments to Draper that added to the town's unique character. From 1950 to 1964, Draper had its own airport when Vern Carter built the Carter Sky Ranch. Carter gave flying lessons at the Sky Ranch before relocating to Lehi. Draper also had a dinosaur roadside attraction for a short time in the 1960s when the sculptor Elbert Porter displayed his life-sized dinosaur statues outside of his studio. Lastly, in 1966, a movie titled *The Devil's Brigade*, starring George Hamilton, was shot at Lone Rock above Corner Canyon.

Draper was incorporated in 1977, a move partly inspired by the fact that the neighboring city of Sandy was continuing to expand south. As suburban sprawl began to increase throughout the Salt Lake Valley in the 1970s, developers began buying up Draper's farmland for housing and businesses.

To say that the past 30 years have transformed the city of Draper would be a vast understatement. Between 1990 and 2020, Draper's population grew from 7,000 to over 51,000 according to the US Census. The trend is in keeping with the remarkable population growth happening along the entire

Wasatch Front. Yet Draper has grown and changed over the past three decades in other ways as well, transforming from a farming community to a hub for tech companies and a cultural destination.

In 2019, *Money* magazine voted Draper as the sixth best place to live in the United States. *Money* referred to Draper's growing business landscape as "a beacon for expanding startups looking for a place to put down roots." As of 2022, the city's median household income was $117,266. Ebay and 1-800-Contacts are just two of the major corporations that have business campuses in the city.

Draper is thriving as a cultural destination. There are multiple venues for homegrown arts and musical entertainment. Draper Historic Theatre has been presenting community theater for decades in an intimate space that was once the town movie theater. The Draper Philharmonic & Choral Society was established in 2017 and presents concerts at venues throughout the Greater Salt Lake Valley. In the summer months, Draper Amphitheater echoes with live music and theatrical productions staged by the Draper Arts Council. Each July, thousands celebrate Pioneer Day with a rodeo, parade, family activities, games, evening concerts, and fireworks at Draper Days and Draper Nights.

In 2014, Loveland Living Planet Aquarium moved from a smaller space in Sandy to a brand new building and grounds in Draper, which attracts over one million visitors annually. The aquarium purchased a huge outdoor stage structure, "The Claw," used by the band U2, and installed it on the grounds. The illuminated structure, now called the Ecosystem Craft & Observatory or EECO, has forever changed Draper's skyline.

Amid all of this growth, however, there has also been a concerted effort on the part of the community to preserve and celebrate Draper's pioneer and agricultural heritage. The Draper Historical Society, founded in 1981, is dedicated to preserving the history of Draper. The society maintains a museum that is open to the public showcasing a collection of artifacts, photographs, and written histories donated by local families. The Draper Historical Society published a three volume set of books on the history of the community. Two of the volumes consist of biographical profiles of Draper's citizens since the time of its founding.

The Draper Historic Preservation Commission works to maintain historic homes. The preservation commission spearheaded the preservation and establishment of the Sorenson Home Museum, which opened to the public in 2003. Other accomplishments of the preservation commission include the moving of the Day dairy barn to Draper Park, the creation of walking and driving tours of historic Draper, and the efforts to save the Park School. The latter was purchased by a developer in 2015, and its classrooms now house a variety of small businesses.

A typewritten sheet in the collection of the Draper Historical Society describes the organization's founding and concludes with anonymous quotes that capture the feelings of many Draper citizens regarding the importance of the city's history. "The past is when memory infuses and enriches the present," reads one quote. The challenge for Draper citizens in the years ahead will be to balance growth with preserving the past.

# One

# THE PIONEER ERA

On July 24, 1847, the first company of pioneer members of The Church of Jesus Christ of Latter-day Saints, led by Church leader Brigham Young, made their way into "New Zion," the northern end of the Salt Lake Valley. Persecuted for their unorthodox beliefs and practices, particularly polygamy, the early Latter-day Saints were violently driven out of the communities they had established earlier in the American Midwest. Seeking land where they could live and worship as they pleased, the Latter-day Saint pioneers pushed into the open western territory then owned by Mexico. They traveled in wagons, on horseback, and on foot, often suffering from hunger, disease, and exposure to the elements along the way.

The settlers established Great Salt Lake City and set about exploring the area beyond. Two men, Joshua Terry and Levi Savage, ventured into a cove referred to as Sivogah or "the willows" by the local Ute and Paiute peoples in the southeastern part of the Salt Lake Valley. The area was subsequently used by the pioneers to herd cattle and horses.

Latter-day Saint pioneers Ebenezer Brown and his wife, Phebe Draper (Palmer) Brown, came to the Salt Lake Valley in 1848 with their large, blended family. They set out to find good land for grazing where they could raise cattle to sell to settlers on their way to the goldfields of California. Ebenezer's son John Weaver Brown wrote in his journal, "We discovered unoccupied land in a large cove in the southeast corner of the Salt Lake Valley, through which the water of four canyons ran."

The Browns built a cabin there and moved to their new home in 1850. The area at that time was called South Willow Creek. Other family members soon followed them, including Phebe's brother William Draper Jr., who became the first presiding church elder. By 1852, there were approximately 20 families living in the area. In October 1853, the townspeople applied for a post office and tried to register their settlement as Brownsville in honor of Ebenezer Brown. As there was already a Brownsville in Utah Territory, which later became Ogden, South Willow Creek was renamed Draperville. As of the 1890 census, the name was shortened to Draper.

Fearing the possibility of Native American uprisings, the settlers decided to construct a fort for protection. Ebenezer Brown donated land near the area where Draper Historic Park is now located, and the construction of the fort walls out of adobe bricks began in 1853. Many Draper families built their cabins near or within the fort. When the threatened uprisings did not occur, the fort was never completed, and there are no physical remains of the walls.

Life was not easy for the people of Draper in the earliest years. Planting and growing crops took time, and at first, the settlers survived by hunting, fishing, and foraging for wild fruits and edible plants such as sego-lily bulbs. Early crops were often lost due to lack of water and invasive pests. Eventually, the settlers dug wells and developed ditch systems and irrigation methods. The construction of canals, reservoirs, and an equitable system of sharing water soon followed.

Their dwellings were primitive at first. People lived in dugouts, log cabins, and converted wagons. Later, they built homes out of adobe bricks. Homes constructed out of fired bricks came later.

Brigham Young sent pioneers with special skills to settlements where they were needed. A blacksmith, a shoemaker, and a teacher came to Draper by the mid-1850s. The first shops were set up in the fort during that decade.

In 1857, when the federal government officially established Utah Territory after the Mexican War, President Buchanan sent a new territorial governor to Salt Lake City. He was accompanied by a 2,500-man military unit lead by a commander named Johnston. Brigham Young and other church leaders interpreted the Army's coming as religious persecution. Martial law was declared, and Young deployed a local militia to delay the troops. Several men from Draper were in that militia, which confronted the federal troops and damaged their supply lines. The Utah War came to an end when the federal government delivered an amnesty proclamation to Young and the other Latter-day Saint leaders in 1858. Utah Territory gained a new governor as well as an army garrison.

When the American Civil War began in 1862, a few Draper settlers left home to fight on the side of the Union. The federal troops who were stationed in Utah Territory left to fight in the Civil War and were not there to defend the settlers when the Black Hawk War broke out in 1865. Territorial conflicts and anger over stolen livestock erupted into a series of deadly confrontations between settlers and Native Americans that lasted several years. Some Draper settlers fought in the conflict, but there is also anecdotal evidence that others shared their food and supplies with Native Americans who had lost their hunting and foraging grounds.

It should be acknowledged that the vast majority of Draper's settlers were white and came from Northern Europe or the eastern United States. However, slavery was legal in Utah Territory, and some Latter-day Saint pioneers from the South brought African American slaves with them. A settler named George Bankhead from Mississippi brought a slave named Nancy—or Hannah by some accounts—and her five sons to Draper. Nancy kept house while her sons worked on Bankhead's farm. After the Emancipation Proclamation in 1865, Nancy and her sons left Draper and settled in Union.

The 1860s were distinguished by many changes to the pioneer community, which transformed it into an established town. A church and community center known as the Old White Meetinghouse was built and used as a school. In 1867, the town elders created city blocks and devised deeds for property. The first co-op store in Draper was set up in 1869.

New communication technology came to Draper in the form of the telegraph, but by far, the greatest change to life in Utah Territory was brought about by the railroad. The Utah Central Railroad was completed to Salt Lake City by 1870 and was extended to Draper in 1871. Many men in Draper went to work on the railroad.

With the completion of the tracks, mail and newspapers were able to reach the people of Utah Territory faster than ever before. Draper citizens could travel to Salt Lake City in a few hours and could go virtually anywhere in the country, for that matter, with greater ease and comfort. More settlers of different faiths and ethnic backgrounds made their way to Utah Territory—and Draper—via rail. The days of grueling journeys to Zion by wagon and handcart were largely over, thus ending the pioneer era.

Titled *Draper Vista*, this painting of the land that became Draper hangs in Draper City Hall. It depicts the area's natural beauty as it appeared in the early days of the pioneers. The painting captures the vast mountains as well as the verdant grasslands that met the first settlers. The presence of sheep in the foreground illustrates the fact that many of Draper's pioneer farmers raised sheep. The mural was painted by Utah artist Linda Jo (Curley) Christensen and was commissioned by the Draper Visual Arts Foundation in 2004. Born in Idaho in 1956, Christensen is an acclaimed painter of Utah landscapes and wildlife. Her work may be seen at The Church of Jesus Christ of Latter-day Saints Conference Center in Salt Lake City and at temples throughout the world. (Photograph by David J. Weinstein; courtesy of Draper Visual Arts Foundation.)

Charles W. Carter came to Salt Lake City from England in 1859 where he became one of the first photographers in Utah. He took many pictures of wagon trains, including this one at Echo Canyon, a 25-mile stretch of the Mormon Trail west of what is now the Wyoming border. (Courtesy of the Church History Library, The Church of Jesus Christ of Latter-day Saints.)

This photograph of two Ute men was taken by Charles W. Carter around the time of the Black Hawk War, the longest conflict between Utah settlers and Native Americans. The Utes were pushed out of their traditional hunting and foraging lands by settlers and sometimes resorted to stealing their livestock. The resulting tensions led to war. (Courtesy of the Church History Library, The Church of Jesus Christ of Latter-day Saints.)

Joshua Terry camped overnight in the area later known as Draper in 1847. He was a mountain man, scout, and "Indian interpreter" for Brigham Young. Terry was born in Canada in 1825 and joined The Church of Jesus Christ of Latter-day Saints in 1840. An early explorer of the South Valley, he later walked all the way to Fort Bridger, Wyoming, where he found work as a hired hand. In the course of his travels, Terry was sometimes treated as a friend and at other times as an enemy by different Native American tribes. His first two wives were Native American women who both died at an early age. Terry moved to Draper in 1856, where he served as justice of the peace for 11 years. He was also on the school board and was elected president of the Draper Irrigation Company in 1888. Terry, photographed here with his third wife, Emma Reid, was the father of 16 children and died at age 90 in 1915.

Ann "Pee-che" Greasewood was a Shoshone woman who was Joshua Terry's second wife. They met and married at Fort Bridger, Wyoming. The couple's children included George Terry, who served as an interpreter and advocate for the Shoshone in their negotiations with the US government. Ann died from tuberculosis in 1857 and is buried in Draper Cemetery.

Built by Joshua Terry around 1879, this brick house has served as both a private home and a business over the years. In 1929, it became the home of farmer and landowner C.H. Carlquist. In the 1990s, the Sundquist family transformed it into Draper's first bed and breakfast. The Charleston Draper, a fine dining establishment, opened there in 2016. (Photograph by LaRayne Day; courtesy of Draper Historic Preservation Commission.)

Ebenezer Brown was already 49 years old when he became the first Latter-day Saint pioneer to settle in the area that is now Draper. The son of a Scotsman who fought in the War of 1812, Brown was born in the state of New York near Syracuse. He became a member of the Church and moved with the Latter-day Saints from Pennsylvania to Missouri to Illinois. When his first wife, Ann, died in Illinois, Brown married a widow, Phebe (Draper) Palmer. Brown joined the "Mormon Battalion" in the Mexican War before coming to the Salt Lake Valley. After settling in the area which came to be known as South Willow Creek, Brown took a second wife, Samantha Pulsipher, and later a third, Mary Wright. Brown and his large family raised cattle and horses and grew the first crops of corn and wheat in South Willow Creek, which they shared with the other settlers. He went on a mission to Nevada in 1858 but returned to Draper. Ebenezer Brown died in 1870 at the age of 69.

Phebe (Draper Palmer) Brown, a widow with six children, married widower Ebenezer Brown in 1842 in Pleasantville, Illinois. Once they were settled in South Willow Creek, Phebe took it upon herself to greet and help new arrivals to the town. As Phebe could read and write, she was the de facto town postmistress. She raised and cared for many children, including those of Ebenezer's other wives who died before her.

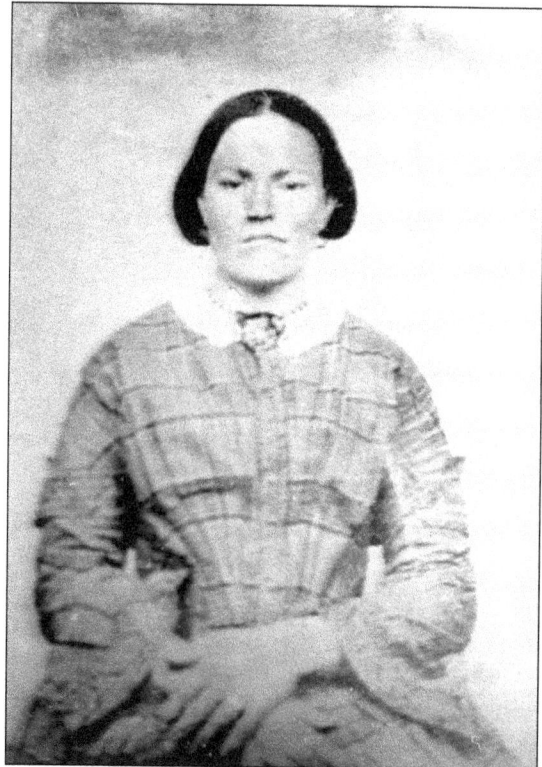

Born in Kirtland, Ohio, Samantha Pulsipher was only in her teens when her family walked across the plains from Missouri to the Salt Lake Valley in 1851. She met Ebenezer Brown in Provo and became his third wife in 1853. When Ebenezer was sent to Nevada to establish a new settlement, Samantha went with him. She was the mother of 10 children and was skilled at spinning and weaving.

William Draper Jr., for whom the town was named, settled first in Millcreek before moving to South Willow Creek in 1850 at the invitation of his sister Phebe and her husband, Ebenezer Brown. The family grew wheat and raised horses and cattle on their L-shaped farm, which was located near the intersection of present-day Pioneer and Fort Streets. Draper and his brother Zemira expanded Willow Creek by diverting several small streams into one to water their crops. William's first wife, Elizabeth (Staker) Draper, was known as "Aunt Betsy" in Draperville and was the town midwife and nurse in addition to raising 11 children of her own. William became the first presiding elder of the small Draperville branch. He left South Willow Creek when Johnston's Army arrived and settled first in Alpine and later in Spanish Fork.

Perry Fitzgerald was born and raised in Pennsylvania. He came to the Salt Lake Valley from Nauvoo, Illinois, on July 24, 1847, as a member of Brigham Young's first company and assisted in raising the first flag on Ensign Peak. Early on, he made a home with his family in Salt Lake City where he helped build the fort, which later became known as Pioneer Park. Fitzgerald then lived for a time in Millcreek before coming to South Willow Creek. He defended the settlers against threats from Native Americans and was elected first lieutenant of the Willow Creek militia. A man with a kindly disposition, he operated the town's first store in one room of his cabin. Fitzgerald raised sheep and cattle and appreciated fine horses. In the course of his life, Fitzgerald had multiple wives with whom he fathered 20 children. He died in 1889 at age 74.

Perry Fitzgerald built this cabin in 1851 along Willow Creek but later moved it due to floodwaters. It had two rooms downstairs with an attic and a fire pit. When the family later moved into a brick home, the cabin was used as an outbuilding on their farm. The Draper Historical Society relocated the cabin to Draper Park in 1990. It was later moved again to 1160 East 12400 South.

Fitzgerald built this home around 1870. It is one of the oldest homes in the southern part of the valley constructed out of fired bricks. Fitzgerald lived here with his third wife, Agnes Wadsworth Fitzgerald, who was known for her skills as a nurse and healer. The house was later used as the office of the Draper Chamber of Commerce. (Photograph by LaRayne Day; courtesy of Draper Historical Preservation Commission.)

19

Orrin Porter Rockwell was a lawman in Utah Territory who served as a bodyguard to both Joseph Smith, founder of the Latter-day Saints movement, and his successor, Brigham Young. He ran an inn and brewery at the Point of the Mountain in Draper. A local hiking trail is named for him. (Courtesy of the Church History Library, The Church of Jesus Christ of Latter-day Saints.)

Born in Kentucky and named for the seventh president of the United States, Andrew Jackson Allen and his family were among the first groups of Latter-day Saint pioneers to arrive in the Salt Lake Valley in 1847. He settled first in Millcreek but moved to South Willow Creek in 1852. Allen joined the militia against Johnston's Army and served under Porter Rockwell.

British immigrant John Boulter helped build the Draper fort and other buildings as a brickmaker and mason but is remembered most as a musician who organized community dances. He assembled the first orchestra in Draper in 1860, which provided music for dances that often lasted into the morning hours. He married Ester Ann Munro in 1855. Ester established a hat shop in the Draper fort and was a talented seamstress.

A young runaway from New Jersey, Isaac Mitton Stewart made his way west working as a farmhand. In 1852, he led a company of fellow Latter-day Saint pioneers across the plains to Utah. Stewart was the first bishop of Draper and served for 34 years. His former home on 700 East still stands. (Courtesy of the Church History Library, The Church of Jesus Christ of Latter-day Saints.)

Now a private residence, Bishop Stewart lived in this 16-room home built from adobe bricks with his three wives and 15 children. Stewart spent many hours here working and reading in his study. As adobe bricks can crumble from exposure to moisture and wind, the home requires a well-maintained coat of paint to preserve the structural integrity of the bricks. (Photograph by LaRayne Day; courtesy of Draper Historical Preservation Commission.)

Absalom Wamsley Smith was born in West Virginia. He moved west and married Amy Downs in 1840. The family built a 12-room home known as the Smith Inn and a farm near State Street in Draper in 1852. Smith served in the Draper bishopric and was a member of the first school board. Old timbers from his barn may be seen today as an arbor at the Draper Public Library.

John Enniss was ordained an elder of the Church of Jesus Christ of Latter-day Saints while still living in England. After marrying Elizabeth Boulter, the couple set sail for America. They settled in Draper in 1852. Enniss sold cider and molasses on his farm and had a keen eye for appraising cattle. He served as a school trustee and was vice president of the Draper Irrigation Company.

The sister of John Boulter, Elisabeth married John Enniss in England before immigrating to the United States in 1849. She endured poor health and the deaths of family members along the way but was determined to get to Zion. She enjoyed entertaining in her home and established the first Sunday school in Draper in 1857 with Ann (Wilson) Fitzgerald and Elizabeth Heward.

This photograph of the John Enniss stockyard illustrates the wide open space in which Draper's pioneers first settled. Later, in addition to crops, the settlers planted different varieties of trees, which soon dotted the landscape. In 1874, Elisabeth Enniss wrote in her diary about the construction of a duck pond on the family farm. The pond, which provided the cattle with a place to drink, was

likely used as a swimming hole by the family's children. Many pioneer farmers also used ponds to soak wagons and buggies as their wooden wheels would shrink and crack in Utah's dry climate. Fed by mountain streams, this pond on the Enniss property was located approximately half a block north of 13200 South and 1251 East.

Family lore has it that Henry Day followed his heart to Utah when he met Leah Rawlins in Nauvoo, Illinois. A carpenter from Maine, Day followed the Latter-day Saint pioneers to the Salt Lake Valley in 1850. The following year he built the second home in Draper, joined the Church and married Rawlins. Day confronted federal troops in the Utah War and later helped to build the railroad in Utah.

In 1862, Henry Day took a second wife, Elizabeth Cottrell, a young pioneer woman from England. They are pictured here later in life with their children. Elizabeth spent her days raising her children as well as those of Henry's other wives when they passed away. She spun yarn and wove it to make clothing and carpets. "Aunt Lizzie" lived to the age of 80.

Built in 1851 from adobe bricks, Henry Eastman Day's home is one of the earliest houses built in Draper. It is still in use today as a private residence. Elias John Day, the fourth son of Henry and Elizabeth Cottrell, started a dairy farm across the road from this house. (Courtesy of LaRayne Day.)

Veterans of the Black Hawk War and their families gathered for a camping trip in Spanish Fork in this undated photograph. The names of most of the people pictured here are unknown; however, Joshua Terry (seated) is seen with his wife, Emma (standing next to him). The Black Hawk War, which began in 1865, was a series of violent skirmishes between Latter-day Saint pioneers and Native Americans.

Lauritz Smith was born in Denmark and became an apprentice blacksmith as a teen. He joined The Church of Jesus Christ of Latter-day Saints in 1851 and was ordained an elder by age 22. Smith was determined to join the Saints in Utah and boarded a ship bound for America. He married his first wife, Maren "Mary" Kristine Mikkelsen, seen here, onboard the ship. The ship landed in New Orleans where Smith worked on the docks until he had saved enough to travel to Kansas City where other Latter-day Saints prepared for the journey west. On their long trek to Utah Territory, Lauritz became ill and Mary drove the team of oxen. The Smiths arrived in Salt Lake City in the fall of 1854 and were sent to Draper by Brigham Young. They raised five children in Draper. With Mary's consent, Lauritz later married Johanne "Hannah" Jensen, with whom he had 12 children.

When Lauritz Smith arrived in Salt Lake City, Brigham Young sent him to Draper as the town needed a blacksmith. Smith, shown here in old age outside his shop, became known for his skill at making plows, and many farmers brought items such as gun barrels and pistols to him to be melted down and made into plowshares.

Thomas Stokes's family moved to Draper in 1860, where they established a farm. Stokes Avenue, East 13450 South, is located on what was part of their land. Stokes brought many Latter-day Saint settlers to the Salt Lake Valley from Wyoming and Missouri in his wagon. He fought in the Black Hawk War and later survived losing his left hand while blasting rock for an irrigation ditch in Willow Creek Canyon.

Mary Allen (Phillips) Terry was the wife of William Reynolds Terry, the first schoolteacher in Draper. In 1859, the Terry family followed Brigham Young's call to settle in St. George. Mary later survived both her husband's death in 1868 as well as a flood that swept through their home. She returned to Draper and is shown here with one of her grandsons.

Born in England in 1831, Thomas Williams immigrated to America and came to Utah Territory in 1855. Williams taught school for a couple of years but left his mark on Draper for the system of irrigation he established. According to Williams's system, farmers owned shares of water depending on the size of their farms and how much they contributed to building and maintaining irrigation infrastructure.

Both immigrants from Denmark, Peter Sorenson and Martina Thompson married in 1880. Peter worked for the railroad, and Martina was an avid gardener and hostess. Their home, as it was originally constructed in 1882, was a one-room adobe dwelling, but they added more rooms over the years to accommodate their 10 children. Martina's brother James Jensen was in the stake presidency. At his invitation, the apostles of the Church went to the Sorenson home for meals when they took the train to Draper during stake conferences. Danish-speaking Latter-day Saints in the community met at the home on Sunday evenings for songs, testimonies, sermons, and prayers in the Danish language. After Peter died in an accident in 1914, Reuben Sorenson moved his family into the house and helped his mother until her death in 1954. The Sorenson home opened to the public as a museum in 2003.

This photograph of the Draper train station may have been taken around the turn of the 20th century, but the Utah Southern Railroad actually completed the line from Salt Lake City to Draper in 1871. The railroad station in Draper was the southernmost point in the rail system before the tracks were extended to Lehi in 1872. New passenger trains on the line afforded settlers the ability to travel from Draper to Salt Lake City and back in one day. The early trains stopped at Sandy, Fort Union, and Murray along the way. The coming of the railroad to Draper also meant increased business opportunities and reduced shipping times for goods. Draper's shopkeepers began purchasing more items for resale. Families were eager to purchase things that had been difficult to ship by wagon such as farm machinery, household goods, cooking stoves, pianos, and organs.

# Two

# AN AGRICULTURAL
# COMMUNITY

The history of agriculture in Draper is distinguished by the wide variety of crops grown there as well as by the hard work and tenacity of the community's farmers. Draper farmers grew everything from grains such as corn and wheat to sugar cane, sugar beets, and all kinds of fruits and vegetables. They raised cattle, sheep, and, most famously, poultry. The Draper community faced challenges such as crop-destroying insects in the early years to shifts in the agricultural marketplace later on. For well over a century, Draper's farmers made their living supplying wool, eggs, milk, meat, and produce to people in neighboring communities and beyond.

In 1850, Norman Brown planted 10 acres of corn, which was the first crop in South Willow Creek. The entire community contributed their water to the crop and shared in the harvest. Other crops such as wheat, oats, potatoes, and sugar cane soon followed. The sugar cane was milled to make molasses.

Throughout Draper's entire history as a farming community, farmers often worked cooperatively. They came together for seasonal tasks such as threshing wheat, oats, and barley; harvesting crops; and mowing hay. Neighbors shared equipment and traded goods and produce.

In the pioneer era, most Utah farmers kept a few head of cattle and sheep. By the turn of the 20th century, the state was a major producer of wool. Some Draper farmers maintained large herds of sheep and contributed substantially to Utah's wool industry.

Early on, the Draper pioneers formed a garden club that encouraged members to grow fruits and vegetables as well as a wide variety of trees that would produce shade and sturdy wood for construction. A few farmers even experimented with making silk by raising silkworms fed with leaves from locally grown mulberry trees. By the 1870s, people from neighboring communities were buying molasses, produce, and poultry from Draper farmers.

All of this required a great deal of water in arid Utah. The pioneer farmers diverted streams and dug irrigation ditches to water crops. Canals were built to consolidate the water coming from mountain streams. The East Jordan Canal, which draws water from the Jordan River, was completed in 1882. At first, farmers owned shares of water depending on the size of their farms and how much they contributed to building and maintaining irrigation infrastructure. However, the issue of having fair access to water was sometimes a contentious issue between neighbors.

Eventually, a group of farmers decided to join forces and incorporate to ensure a fair system of water rights. In 1888, the Draper Irrigation Company was founded. After paying fees, each stockholder received the right to use the company's water. Over the next several decades, the Draper Irrigation Company oversaw upgrades to Draper's water infrastructure for both farms and homes. Pipes bringing culinary water into homes were first installed around the turn of the 20th century.

The presence of a modern source of clean water in addition to electricity, which came to Draper around 1912, allowed for the establishment of the large-scale poultry industry for which the town became famous. In 1918, a group of eight Draper farmers set up a cooperative to buy chicken feed and market eggs. Eight years later, the cooperative was registered as Draper Poultrymen Incorporated.

Many of the farming families in Draper did not have enough land to generate an income from farming alone, but chicken farming did not require much space. At the time of the 1930 census, one third of all farmers in Draper were in the poultry business. The Utah Poultry Producers Co-op Association, which was affiliated with the Utah State Farm Bureau, opened a branch in Draper in 1933 and acted as a supplier and distributor for local egg farmers. The Draper Egg Producers Association was organized around the same time as a competitor. Draper eggs were soon sold across the nation, and the town became known as the "Egg Basket of Utah."

During World War II, Draper Poultrymen shipped thousands of cases of eggs to servicemen all over the world. Veteran George Sorenson recalled that seeing an egg crate labeled "Draper Egg Producers—The Egg Basket of Utah" outside of a mess hall on Corregidor Island brought homesick tears to his eyes. "The egg plant was only one block east of the home where I grew up," Sorenson wrote.

Dairy farming was another major part of Draper's agricultural economy. Throughout the 20th century, Draper farmers produced milk for dairies in Salt Lake City, Ogden, and beyond. Transporting milk to the dairies where it could be bottled and sold or processed into products like butter and cheese on a mass scale was difficult in the days before refrigeration. In 1924, a cooler where farmers could bring their milk was built at the corner of 700 East and 12300 South. In 1950, the Utah State Board of Health began requiring all dairy farms to install cooling tanks as part of their milking equipment. Modern refrigeration and trucking allowed Draper farmers to transport and market their milk all along the Wasatch Front.

For a time, sugar beets were a major crop in Utah. In the pioneer years, before the process to derive sugar from sugar beets was fully developed, settlers used molasses as a sweetener or imported white sugar from back East. An inventor named E.H. Dyer built the first successful sugar beet-processing factory in California in 1879. He built another in Lehi, Utah, in 1891, which inspired area farmers to take up sugar beet farming in that decade. By 1932, Draper had 500 acres of sugar beet crops. Draper farmers sent their sugar beets to the West Jordan sugar factory for processing.

Older children and teens worked on thinning the crops over summer break. In the fall, classes were suspended for two weeks so that the students could help harvest the beets. To this day, students at Jordan High School in Sandy are known as "Beetdiggers."

The decades following World War II brought new challenges to Draper farmers. The demand for Utah-grown sugar beets diminished as technological advances made sugar processing by other means faster and more efficient. At the same time, by the late 1950s, Draper poultry farmers faced greater competition from egg producers in California. In the face of changes such as these, several Draper farmers turned to mink farming during the 1960s. Draper farmers also continued to grow produce that was sold both locally and to national canning companies.

There are many factors that led to the gradual disappearance of family farms in Draper beginning in the 1970s and 1980s. Small farms across the entire US faced greater financial pressure with the rise of corporate industrial farming. The growing population along the Wasatch Front demanded more space for housing and infrastructure. In some instances, members of the younger generation in Draper left family farms to pursue other careers.

As the 21st century began, more and more of Draper's farmland was developed for housing, businesses, schools, and medical facilities. Today, just a few farms remain in Draper, although many citizens wish to preserve the agrarian character of the community.

Pioneer photographer Charles W. Carter took this photograph of dead grasshoppers. Clouds of the insects could blot out the sun, and they would devour everything green in their path. In the days before pesticides, farmers would try to eradicate the grasshoppers by driving them into straw-filled ditches that were then set on fire. (Courtesy of the Church History Library, The Church of Jesus Christ of Latter-day Saints.)

In this photograph from 1905, a group of farmers operates a threshing machine owned by the Jensen family of Crescent. Threshing machines harvested grain by shaking the seed kernels from their stems, which were fed into the machine. Threshing was a communal event in which farmers often shared one machine, assigned specific jobs to each person, and usually finished the massive project in one day.

Horses, cattle, sheep, and goats all consume hay, especially during winter or times of drought when they cannot graze. A farming community like Draper, therefore, required tons of hay. Harvesting and storing hay was one of the most time consuming farm tasks until the invention of the hay baler, which came into use in Draper by the 1940s.

James Green was born in England in 1837, immigrated to Utah Territory, and married Ellen Agnes Draper, a niece of William Draper Jr. and Phebe (Draper) Brown. Green is seen here driving a horse-drawn wagon. Although automobiles were introduced to Utah at the turn of the 20th century, horse-drawn wagons could still be found on Draper's farms throughout much of the 1900s.

36

Willard Boulter Enniss, born in South Willow Creek in 1857, played a significant role in increasing Draper's water supply. He was involved with the incorporation of the Draper Irrigation Company and served as a record keeper. Enniss helped settle water rights cases and was instrumental in the location and design of a dam constructed at the head of Bell's Canyon.

Known as The Castle, this home was originally built by W.B. Enniss and his wife on their homestead property in 1898. In 1950, it was purchased by the Japanese American Akagi family who grew strawberries and many other fruits and vegetables on their Draper farm. As of 2022, the home is still a private residence.

Born in Draper in 1869, Jackson Rial Allen was a sheep farmer, a teacher, a veterinarian, a visionary, and an entrepreneur. In this photograph, Allen posed at his home with several of his purebred Cotswold sheep. He founded Excelsior Livestock Farms with his three brothers and worked to upgrade stock quality in the area. Among his many accomplishments, Allen served on the Board of Canal Presidents and the Jordan District Board of Education. He was president of the Draper Irrigation Company, worked to upgrade Draper Park, and was instrumental in bringing electric power to Draper. His home, which was built in 1900, was among the first in Draper to be wired for electricity. It is one of the few surviving residences designed by architect Richard K. A. Kletting, who also designed the original main building at Saltair. (Used by permission, Utah State Historical Society.)

This photograph was taken around 1910 and likely depicts the children of J.R. Allen and Matilda Day Allen. Children who grew up on farms at that time played and had fun but were also expected to do many chores. Benjamin Rial Allen recalled milking cows, feeding pigs and chickens, gathering eggs, and bringing in wood and coal for the stove as a child.

Before the days of modern refrigeration, farmers maintained icehouses such as this one on the J.R. Allen farm. Ice was cut from frozen ponds in the wintertime using handsaws and dragged by sled to be stored in the icehouse. Chunks of ice packed in sawdust or straw would last all summer and could be used to keep foods cold. (Courtesy of Paul Smith.)

During the first decades of the 20th century, the Allen family managed approximately 8,500 head of sheep, which were raised primarily for wool. These sheep were awaiting the "sheep dip," a bath where their wool would be treated with a special solution to prevent tick and mite infestation. (Courtesy of Paul Smith.)

Members of the Allen family were photographed at sheep camp in the early 1920s. Mary Day (Allen) Steadman remembered that the whole family would camp with the herd for a month each summer. The children pitched in to help set up camp, build corrals, and dip the sheep. At night, they slept on pine boughs with lots of blankets to keep warm in the chilly mountain air.

Horses were essential to managing large herds of sheep on the range. To this day, horses allow shepherds to travel over rangelands quickly to keep up with livestock on the move and to help move herds from field to field. On the farm, horses pulled everything from wagons to plows and other machinery necessary for harvesting crops. (Courtesy of Paul Smith.)

Having grown a thick coat of wool over the winter, sheep were rounded up for shearing in the spring. A skilled farm worker could shear a sheep by hand in approximately 15 minutes. These workers posed for a photograph sitting atop sacks of wool. Wool from Draper's sheep herds was sent to Salt Lake City for processing.

In 1917, 12300 South was a quiet country road with farmland on either side. The telephone poles were a relatively new addition. Long-distance telephone service between Salt Lake City and Park City was established in 1883 with lines installed as far south as Nephi by the 1890s. The first transcontinental telephone line was created in 1914 when the final telephone pole was erected in Wendover, Utah.

For centuries, farmers the world over have had barn cats to keep the rodent population in check, and Draper's Sorenson family was no exception. The Sorensons had a small orchard and raised a few chickens, turkeys, and cows. Their home was opened to the public as a museum in 2003.

First developed in the late 1860s, steam tractors were a mainstay of American farms until they began to be replaced in the 1920s by smaller tractors with internal combustion engines. This Case steam tractor is being used for threshing grain. Threshing involved many tasks: pitching bundles of grain stalks into the thresher, supplying water for the steam engine, hauling away the freshly threshed grain, and storing it in a granary.

Leona (Allen) Smith carefully picked gooseberries in the early 1910s as the bushes have sharp thorns. Many Draper families were subsistence farmers, and no source of food was ignored or went to waste. Two varieties of gooseberries are native to Utah. The tart berries may be eaten raw or made into jam or pies. (Courtesy of Paul Smith.)

43

This aerial photograph of the Mickelsen's and Parkin's chicken farms from the early 1930s shows the scope of the operation. The Mickelsen brothers headed the Draper Egg Producers Association at the time of its founding. Relius Mickelsen started out in the poultry business with only 32 hens in 1921. The coops shown here housed approximately 15,000 chickens. The Parkin family sold their eggs to the Draper Egg Producers Association as well. Draper farmers raised White Leghorn chickens, which are known as excellent egg-layers. Each hen produces an average of 280 to 300 eggs per year. This photograph also shows the Lewis McGuire farm in the background. The McGuire farm had been established in the 1860s by John William and Tamar (Stokes) McGuire. Their son Lewis was the last of his family to live on the old farm where both dairy cows and chickens were raised.

This advertisement for Draper Poultrymen Inc. encouraged would-be chicken farmers to move to Draper and highlighted the benefits of working as part of a cooperative. A clever marketer created the image of chickens spelling out "Draper Utah" by "writing" the huge letters on the ground with chicken feed.

When Doris (Burgon) Smith's husband, Herman James Smith, died in an accident in 1949, Doris was left to care for six children, a herd of dairy cows, and approximately 2,000 chickens. She sold the cows and farm but kept the chickens. Doris kept her family afloat by taking on jobs outside the home, and her children helped to collect and prepare eggs for sale. (Courtesy of Helene Terry.)

Raymond Orestes Baker, pictured here second from the right, taught typing and business classes at Draper Junior High and was also a farmer and avid hunter. The Baker family took many hunting trips with friends, such as this one to Beaver, Utah, in 1938. R.O. Baker had a large farm at the mouth of Corner Canyon where he raised dairy cows and a variety of produce.

This photograph of the Perry Fitzgerald barnyard, taken in the 1940s, illustrates the contrast between old and new on the family farm. The old cabin from pioneer days was used as an outbuilding on the farm at the time this photograph was taken. During the 1940s, many Draper farmers began to use hay balers, and the neat bales of hay are visible in the background.

The winter of 1948-49 was one for the history books, as seen in this photograph of Fort Street. A "Corner Canyon Special" snowstorm brought drifts that were as high as telephone wires with snow piled 20 feet deep in places. Horse-drawn sleighs were used by the poultry farmers to haul eggs and feed over the snow.

Plain View Farm, a successful dairy farm, was established by Swedish immigrant Andrew Sjoblom near what is now the northwest corner of I-15 and 12300 South. For years the Sjoblom family sold milk to Clover Leaf Dairy. The barn, with its distinctive green roof, was a Draper landmark. Parties and basketball games were held in the barn's hay loft when it was not in use. (Courtesy of Lorna (Sjoblom) Toone.)

Part of the Day Dairy Farm, this barn was built by Elias John Day in 1922. When the Day family relocated their farm to Payson in 2007, theirs was the last working dairy farm in Salt Lake County. The Draper Historic Preservation Commission moved the barn to Draper Park where it was renovated and is now used as an event space. (Photograph by LaRayne Day; courtesy of Draper Historical Preservation Commission.)

In spite of the Depression, Harmon Eastman Day transformed his family's struggling dairy farm into a highly successful one with a herd of registered Holstein cattle. Three of his children, Norma, Douglas, and James, are seen here with a bull calf around 1933. All of the Day children learned how to milk the cows, pile hay, and ride horses. (Courtesy of LaRayne Day.)

Jack Day posed for a photographer on his family farm in the 1950s, demonstrating how milk was poured into a cooling tank. As of 1950, the Utah State Board of Health mandated that all dairy farms install milk-cooling tanks as part of their milking equipment. (Courtesy of LaRayne Day.)

Brothers Jack and Henry Day bought the family dairy farm from their mother in 1965 and made many improvements to modernize it. Jack took great pride in the herds of registered Holsteins. A few years after this photograph was taken, Jack and Henry received the Distinguished Holstein Breeder of the Year award at the Utah Dairy Convention in 1987. (Courtesy of LaRayne Day.)

These sugar beet digging tools are in the collection of the Draper Historical Society. Farmworkers used them to lift a sugar beet from the soil and then "top" it, cutting the stem and leaves off. The discarded sugar beet greens were left in the field for livestock to eat. (Photograph by David J. Weinstein.)

Jimmy Day and Grif Kimball operated a sugar beet combine harvester on the Kimball farm in 1953. The harvester collected the beets, cut the tops off, and deposited the harvested beets into a separate trailer via a conveyer belt system. The harvesters transformed the sugar beet industry. (Courtesy of LaRayne Day.)

Northrop Enniss Garfield and his wife, Ann, established a thriving farm in 1930 where they grew berries, peas, melons, grapes, and tomatoes. The family also had fruit orchards and raised chickens. In the early 1950s, they built a second home, shown here, on their property located near 1300 East and 13400 South. The Garfield children and grandchildren enjoyed monthly hayrides during family home evenings. Earl Garfield is seen here driving the hay wagon. For members of The Church of Jesus Christ of Latter-day Saints, family home evening is one night per week during which families are encouraged to spend time together doing enjoyable activities and studying the gospel. (Both, courtesy of Anne (Garfield) Covington.)

In 1932, Arthur Eugene Smith purchased a piece of farmland in Draper that included a dewberry patch from his father, Joseph Michael Smith. For years, members of the Smith family would get up at dawn to pick dewberries, which they sold at the farmer's market in Salt Lake City. When Arthur died from a stroke in 1955, his wife, Lucile, became solely responsible for managing the small farm and raising their children. Lucile (Tolman) Smith, seen here looking up from her work in the dewberry patch in the 1950s, was a hardworking farmer's wife. It was said that she was rarely seen outdoors without a hoe in her hand. Dewberries are similar to blackberries but grow on long, thorny vines, which makes picking them difficult. They can be eaten raw but were more commonly made into jam. Lucile sold dewberries by the case and often had a waiting list of people wanting to buy them. (Courtesy of Esther Kinder.)

Bill Marcovecchio and his family moved to Draper in 1958 and established Green Briar Farms at 178 West 12300 South. Marcovecchio is seen here operating an onion loader. The family grew dry onions, cabbage, corn, and cauliflower, among other produce, on acreage around Draper and sold the vegetables to local supermarkets. The onions were sold as far away as the eastern part of the United States. (Courtesy of Paula Summers.)

Marcovecchio prided himself on his even rows of red and green cabbages, seen here in a view from the family's backyard. This photograph was taken before the freeway came to Draper in the 1970s. He sold his land to Utah Retirement Systems (URS) in 1979 but kept farming through the 1990s when URS sought to develop the property. (Courtesy of Paula Summers.)

With the ending of the Vietnam War, Watergate, inflation, and recession, the 1970s were a turbulent time in US history. The decade also brought lasting changes to Draper. The new I-15 Freeway was completed from 13800 South in Draper to Salt Lake City, making the commute north faster than ever. With the growth of new housing developments underway, Draper's water infrastructure was given an overhaul, and a new firehouse was built. Most significantly, the town was incorporated and became Draper City in 1977. Kenneth Hisatake became the first mayor, and the first city council members were elected and sworn in to office the following year. This aerial view of Draper was taken in 1977. The photograph shows the extent to which much of the city was still comprised of farmland. Economic pressures and population growth gradually led to a decrease in the number of family farms in Draper over the ensuing decades.

# *Three*

# DRAPER BUSINESSES OF YESTERYEAR

Throughout much of its history, Draper was primarily an agricultural community. However, business and industry played a significant role in the town as well. Over the years, enterprising citizens stepped up to establish businesses that provided goods and services. Draper supported agriculture-related businesses such as canneries and its renowned poultry industry. The town also utilized its natural resources as the site of silica and sand-mining operations.

The first cooperative store in Draperville, established in 1869, was supplied with goods from the Zion's Co-operative Mercantile Institution (ZCMI) in Salt Lake City. Organized in 1868, ZCMI was founded by church leaders as a store where Latter-day Saints could purchase everything they might need at fair prices. ZCMI sold clothing, shoes, household supplies, furniture, fabric, and more. It became known as America's first department store and supplied co-op stores throughout the valley.

Around 1881, David O. Rideout opened a general store in Draperville, also supplied with ZCMI merchandise. The Rideout Store sold locally grown produce in addition to household goods and farming implements. Its second floor, known as Rideout Hall, became the center of social life in Draper for over 30 years. Built in 1883, Rideout Hall was the site of countless dances and theatrical performances.

Other general stores soon followed. The Draper Mercantile and Manufacturing General Merchandise Store, which came to be known as the "M&M," opened in 1898. There, Draper farmers traded produce, butter, eggs, and more for store goods or were paid for their farm-grown wares with scrip, which could be used as money at the store. Store manager Soren Rasmussen was known for being generous at the candy counter where children bringing butter and eggs to trade could often get a treat.

Edward Miller Rasmussen operated butcher shops at the Rideout Store and M&M store before opening his own grocery store at the corner of 900 East and Pioneer Road in the 1920s. Miller Rasmussen was an innovator. Besides delivering meats and groceries to customers in his Model T Ford truck, he opened a soda fountain and lunch counter in the store. Rasmussen's was one of the first businesses in Utah to have a soft-serve ice-cream machine shortly after it was invented in 1930.

The lunch counter at Rasmussen's was one of the first dine-in eateries in Draper. Earlier in the town's history, Porter Rockwell had operated an inn and pub near Point of the Mountain catering largely to Pony Express riders. The Dunyon family later purchased it and operated Our Mountain Home as an inn for stagecoach travelers through the 1870s. After that, there were no restaurants—or saloons, for that matter—in Draper for many years. Turn-of-the-century Draper had a very different character from neighboring Sandy, which had saloons offering meals, beer, and whiskey to a clientele largely comprised of miners.

Rasmussen's became a popular spot for Draper folks. A few other small cafés that offered simple meals were established from the 1930s through the 1960s in Draper. These eateries were sometimes attached to grocery stores and service stations.

The rise of the automobile and roads to accommodate them in the early 20th century brought new businesses to Draper. Years before the construction of I-15, US Highway 89, known locally as State Street, was the main road from downtown Salt Lake City to Point of the Mountain and Utah County. Gas stations and auto repair businesses proliferated in Draper, particularly around the intersection of State Street and 12300 South, which was known as Draper Crossroads.

By the late 1930s, Draper had a thriving community of small businesses. On Fort Street, S.J. Mickelsen's hardware and lumber store, established in 1912, was busily supplying lumber for the construction of chicken coops. A building housing a small movie theater, drugstore, and doctor's office was constructed just north of the Rasmussen store on 900 East in 1938. Draper had automobile dealerships and auto repair shops, gas stations, and grocery and general stores. There were barbershops, dressmakers, and ice-cream parlors.

Draper soon put into place the means to protect its citizens and their homes and businesses. The town's first sheriff, Harry Nichols, assumed the office in 1930. Draper's volunteer fire department was established in 1947.

Almost everyone in town was associated with the poultry industry from the 1930s through the 1950s. While farmers raised the chickens, others worked at the Draper Poultry plant doing a variety of jobs from candling eggs and repairing feed sacks to marketing and distribution.

Another significant industry in 20th-century Draper was mining. Ancient Lake Bonneville left behind valuable natural resources in Draper. In the 1920s, the Rideout silica mine was located north of Point of the Mountain and shipped silica all over the country via rail. Silica was used in the construction of fireplaces and chimneys. Burt Andrus owned a coal yard and managed a sandpit that provided sand for use on the railroad. Sand and gravel businesses remain in Draper to this day.

The construction of the Utah State Prison also contributed to the local economy. As the population of Utah expanded, the prison outgrew its old space in the Sugarhouse neighborhood of Salt Lake City. Draper was viewed as a desirable location for a new prison because of its remoteness. The State of Utah began construction on the prison near Point of the Mountain in 1940. Until its closure in 2022, the prison provided jobs for some Draper citizens.

The city of Draper as it appears today began to take shape in the 1960s when the first farm properties were sold for development. In 1963, Murray Smith, Griffith Kimball, Evan Hanson, and Earl Toone opened Draper Realty in the renovated old milk cooler building. Draper Realty developed eight properties in subdivisions for homes in the 1960s and 1970s

Land previously used for farming increasingly began to be developed for commercial purposes as well as housing. One early example was the Hidden Valley Golf and Country Club, which purchased the Heber A. Smith farm. When Ethel Carlquist passed away in 1976, a portion of her land eventually became the Hidden Valley Shopping Center.

The incorporation of Draper into a city in 1977 came about in no small part because the townspeople were deeply concerned about the rising tide of development and the loss of community identity. People wanted to have a say in the regulation of growth. There was also worry that the city of Sandy, which had incorporated Crescent, would encroach further into Draper. Two thirds of Draper's registered voters approved incorporation. Lawyer Kenneth Hisatake, who had publicly expressed his concerns, was elected as Draper's first mayor in February 1978.

A harbinger of Draper's future in the tech industry arrived in 1978 when Applied Digital Data Systems, a video display computer terminal company, opened on Frontage Road. Many hoped that it would bring sorely needed jobs to Draper, but the venture was short-lived. It was not until after the launch of the World Wide Web and the rise of e-commerce in the 1990s that tech companies like eBay came to Draper. Today, the city is considered an integral part of Utah's "Silicon Slopes."

The work of the blacksmith was essential to farming communities. Niels Boberg established his blacksmith shop on the southeast corner of 12300 South and 900 East around 1880. Boberg and his wife, Caroline, originally from Sweden and Denmark, respectively, first settled in Millcreek after making the long journey to Utah Territory. They came to Draper in 1867 upon hearing that Lauritz Smith was seeking an apprentice in his blacksmith shop. Trained by Smith, Boberg became known for his skill at making cattle-branding irons and tools. Although the bulk of his work was shoeing horses, Boberg also made metal rims for wagon and buggy wheels and fashioned more intricate items such as fireplace tongs, nutcrackers, and locks at his shop. Using a massive pair of tongs, Boberg worked by heating metal until it was white hot in the forge before placing it on his anvil. Sparks would fly as he hammered the metal into its desired shape. Draper's last blacksmith shops closed in the 1950s.

The Rideout Store was a popular destination in Draper not only as a general store but for its upstairs hall where dances were held and theatrical productions took place. This photograph from the turn of the 20th century may depict a local family or members of a traveling theater company who came to town in horse-drawn buggies and wagons.

The mercantile stores that David O. Rideout established in Draper, Sandy, Bingham, and Mammoth were just one aspect of his involvement in the local economy. He helped to establish Draper Irrigation Company and the Draper Canning Company along with interests in mining and farming. Rideout served on the school board for over a decade and was elected as a state senator to the first Utah Legislature.

Several prominent Draper citizens established the Draper Mercantile and Manufacturing General Merchandise Store in 1898 to sell locally produced goods. The building as pictured here was located on the southwest corner of Fort Street and 12600 South. The store, which came to be known as the "M&M," sold everything from clothing, shoes, and fine china to dairy products, fruits, and vegetables. The M&M accepted both cash and scrip, which customers received in exchange for their home-grown produce. A bishop's office was part of the store, which allowed Latter-day Saints to pay their tithing on the premises. Livestock, grain, hay, eggs, and butter were among the goods that would be given as tithing.

Born in Draper in 1885, Joseph E. Mickelsen was a young man when he was photographed with his friend Soren Rasmussen. The photograph was perhaps taken around the time that Mickelsen was called to serve a mission in Norway. Upon returning to Utah, he moved to Idaho with his brothers Orson and Relius to manage the Deseret Sheep Company. In the 1920s, the brothers came back to Draper where they soon founded the Draper Egg Producers Association. They owned the business until 1963, when they sold it to Intermountain Farmers of America. Soren Rasmussen immigrated to Utah from Denmark in 1885 and was a prominent elder in the community. The son of a merchant, he managed the old Draper Co-op store and became the manager of the M&M store in 1901. Rasmussen was known for his generous nature and for giving out candy when customers came to pay off their debts or trade butter and eggs.

In the 1910s, the Draper Canning Company was located near 12349 South 970 East. Many young women in Draper earned money for school clothes by working at the cannery. There were a few canneries in Draper over the years. Factory canning was a major agricultural industry in Utah until local brands of canned foods disappeared with the rise of national brands and mass-produced frozen vegetables in the 1950s.

This photograph of the Andrus Sandpit at 1000 East 12300 South shows part of the conveyor belt system that moved sand from the pit to railroad cars where it was sold for use as engine sand for the Union Pacific Railroad. The sand pit was managed by Millard Burgess "Burt" Andrus, who also operated a coal yard nearby. In 1920, the Carlquist family acquired the property.

Originally constructed in 1912 as S.J. Mickelsen's hardware and lumber store, this building on 12600 South Fort Street has seen many changes over the past century. It has housed Draper's post office, barbershops, a beauty salon, a doctor's office, and many other businesses. (Photograph by David J. Weinstein.)

When Dave's Barber Shop was renovated in 2017, the old wall that used to be part of S.J. Mickelsen's Hardware store was revealed. Preserved on the wall was a poster advertising a rodeo that likely dates to the 1950s. Dave's Barber Shop was established in the space previously occupied by Rex's Barber Shop. (Photograph by David Bean.)

The many cars parked outside of the Draper Poultrymen and Egg Producers Association Plant, seen here in a photograph taken in the late 1920s or early 1930s, hint at the activity within. The facility was comprised of mills where chicken feed was prepared as well as grain silo bins, warehouses for packing and candling eggs and storing equipment, offices for management, and a store. The complex was owned by both Draper Poultrymen Inc. and the Draper Egg Producers Association. The structure started out as a simple warehouse built in 1920. In 1926, Draper Poultrymen Inc. converted it into a poultry feed and processing plant. The retail store was added in 1931, followed by a mill in 1945. The grain bins were constructed in the late 1940s.

From left to right, Mary Fratto, Lois Washburn, Mary Steadman, and another unidentified woman were photographed working in the feed sack department of the Draper Poultry feed plant in 1941. To save money and materials, chicken feed sacks were cleaned, repaired as needed, and reused. Mary Steadman worked at Draper Poultry for one year before becoming the manager of the sack department.

Employees of the Draper Poultry feed plant posed for a photograph outside of the facility in 1941. Some of them had weathered the Great Depression with the company. During those years, the feed plant rotated their employees by scheduling them to work every other week to avoid laying anyone off.

This undated photograph shows a view of the Draper Poultry feed plant's grain storage silos. In the early 1960s, its name was changed to the Intermountain Farmers Association (IFA). Part of the facility caught fire in the spring of 1967 and burned to the ground, but it was soon rebuilt. Today, the IFA has feed mills and retail stores throughout the western United States.

From left to right, Dale Stringfellow, Eugene H. Ballard, Ross D. Ballard, and Art Witherall were photographed while candling eggs at Ballard Egg & Feed. Founded after Eugene and Ross returned home from World War II, the business lasted 20 years. The Ballards purchased the building from Fort Douglas and moved it to Draper. In 1997, Pirate O's Gourmet Grocery moved into the space. (Courtesy of Kathleen (Ballard) Day.)

The automobile transformed life in Draper. In the first decades of the 20th century, however, many roads in the town were not paved. Early motorists wore dusters, hats, and even goggles to combat the clouds of dust. From left to right, P.A. Nielsen, Olivia Meek, Olivia J. Nielsen, Don Meek, and Ben Meek posed in their driving gear with their new automobile sometime in the 1910s.

As traffic increased along Highway 89 at the Riverton and Draper Crossroads in the 1930s, so did the number of service stations offering gas, oil, new tires, and tire repairs along with sodas and ice cream. Bert L. Smith was hired to operate the State Line Service Station in 1932. He later established his own gas station and small grocery store at 1300 East 12400 South.

Henry S. Day, pictured here, was born in Draper in 1913. He studied auto mechanics at Jordan High School and worked with his father at the Poultry City Service Station at the corner of 12300 South and 900 East, pictured here around 1930. It was a popular location with customers coming to fill their tanks on their way in or out of Draper. Day married Effie Blackner in 1931 and purchased the service station one year later. He turned it into an auto repair shop with used cars for sale. In 1942, Day established Henry S. Day Studebaker and built a showroom and shop. He purchased a Ford dealership, Midvale Motors, in 1954 and moved his business to Midvale. Day moved his Ford dealership once again in 1967 to Redwood Road where it remained for many years and became well-known in the Salt Lake Valley. (Courtesy of Tonya Atkinson.)

Miller Rasmussen's store, located at 900 East and Pioneer Road, sold gas, groceries, meat, and produce and also had a lunch counter and soda fountain. In 1935, Rasmussen's became a member of the Intermountain Grocers Association (IGA). At that time, milk cost 25¢ per quart, coffee was 25¢ per pound, and fresh ground hamburger sold for 19¢ a pound.

This photograph shows the interior of Rasmussen's during the 1930s. Pictured from left to right are Richard Orgill, Sonne Rasmussen, a deliveryman, Agnes Fitzgerald Allen, Barbara Rasmussen Fitzgerald, and Milo Rasmussen. Rasmussen's was open seven days a week. The soda fountain was a popular spot to grab a glass of Coca-Cola or a banana split. Students from the Draper Park School across the street would come for lunches and snacks.

Miller Rasmussen's son Sonne remodeled his father's grocery store, giving it a new, modern look when he took it over in 1952. In the long run, however, the small independent store could not keep up with competition from larger supermarket chains like Smith's, and the store closed on New Year's Eve in 1966. The building stood empty at the time this photograph was taken in the early 1970s.

In 1986, Ingemann and Clint Bendtsen acquired the old Rasmussen's/Sonne's Market. They remodeled the exterior and opened the Clog Corner, where they made and sold specialty shoes and leather goods. As of 2022, the exterior of the building has changed little in its appearance since the 1980s and still features the steep gables outside the main doors.

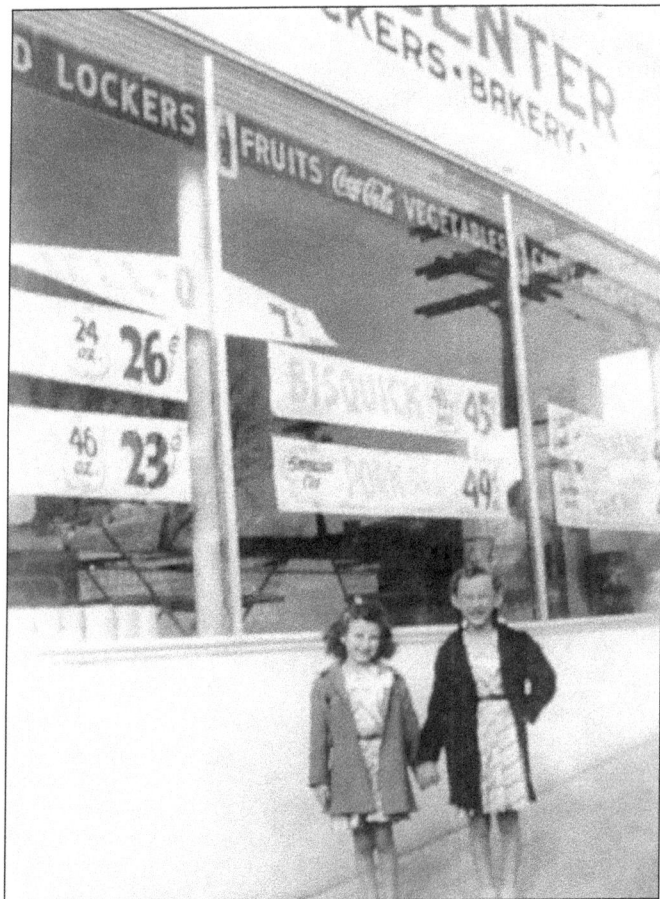

Before home freezers were widely available, people who wanted to keep food frozen for later use had to store it in a commercial frozen food locker. Until 1941, people in Draper had to travel to Sandy to find a locker. This changed when the Frozen Food Center was established on the northwest corner of 12300 South and 900 East. Jack Leslie Brimhall owned the frozen food center, which also sold meats, groceries, and baked goods. Jack and his wife had eight children; two of his daughters were photographed in front of the store. The Frozen Food Center later became known as Frank's Food Town. Today, the building houses a bike shop.

Iona Nelson managed the East Draper Cash Grocery store. At first, it was one of the many service stations and small grocery stores that began to proliferate in Draper in the 1930s as more people owned automobiles and trucks. The gas pumps pictured here used a hand-operated pump to move gasoline from the storage tank into the clear glass cylinder at the top of the pump. A valve on the pump would then be opened to allow gravity to feed gasoline into a vehicle's gas tank. The grocery store became known as a "beer joint" in the late 1930s. The construction of the Draper Alpine Tunnel through Traverse Mountain, which brought water from Deer Creek Reservoir to a water treatment plant at Little Cottonwood Canyon, began in 1938. The East Draper Cash Grocery was transformed into a bar when the men working on the tunnel looked for a place in Draper to unwind with a cold beer. Locals jokingly referred to it as Draper's "Third Ward."

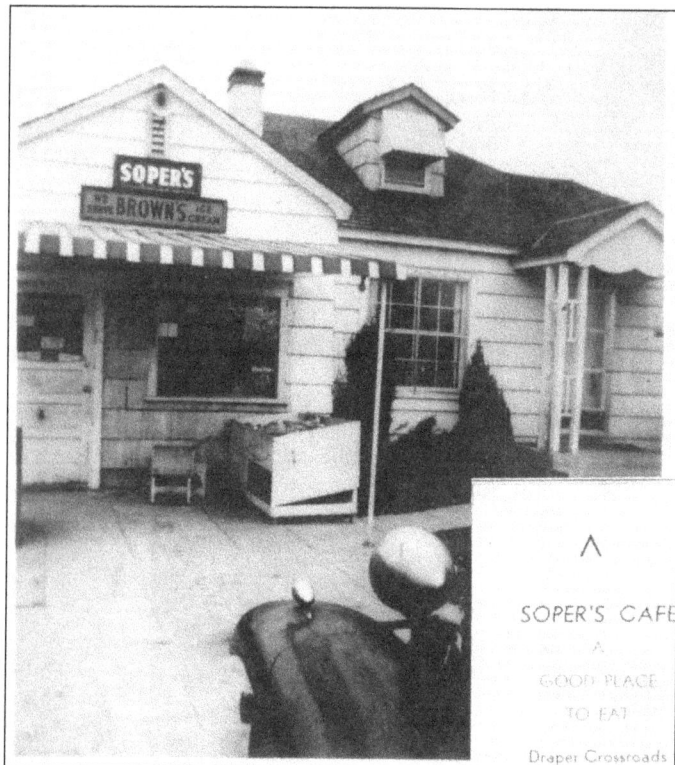

John and Thelma Soper operated a small café owned by Utah Oil at Draper Crossroads after they moved to Draper in 1942. They also established a small grocery store in the garage space at their home on the southwest corner of 12300 South and State Street. John Soper greeted everyone with a warm "good morning" regardless of the time of day. Soper's Grocery remained in business until 1967.

Anne (Garfield) Covington worked as a waitress at Bud's Café, also known as the Ride Out Inn, when this photograph was taken in 1948. The café was located on 12400 South across from the Park School building. Hamburgers, hot dogs, and ice cream were the main items on the menu, and there was a candy counter as well. (Courtesy Anne (Garfield) Covington.)

The Salt Lake County Fire Department was established in the fall of 1921. Twenty-six years later, Fire Chief "Chick" Clay authorized Salt Lake County deputy sheriff James Rayburn Dow to formally organize a fire department in Draper. Dow recruited a team of 10 volunteer firefighters. Draper's first fire station was in a building that had formerly served as a morgue at the Army Air Forces base in Kearns during World War II. Salt Lake County Fire Station No. 5 was dedicated on March 2, 1948. Fire Chief Dow, with Assistant Chief Bill Day and some of the volunteer firefighters, were photographed with the original Dodge fire truck in front of the building. Pictured from left to right are (first row) Bill Day (driver's seat), Harry Ballard, Oral Molyneux, Phil Whetman, and Fire Chief Dow; (second row) Allen "Scotty" Terry and Burt Nichols.

In 1931, George Edgar Whetman purchased the old M&M store building and converted it into an auto dealership with apartments above. From 1936 to 1971, Whetman's was a Ford dealership. This photograph was taken after the dealership was closed. As of 2022, the building is still in use and houses multiple offices.

Swedish immigrant Al Engstrom established an automobile repair shop at the corner of 12300 South and 700 East in 1927. Until his death in 1966, Engstrom was known for his quality work as a mechanic and for his fairness with customers who had trouble paying their bill. Van Terry was photographed in front of Al's Garage with a couple of Mercury Coupes in the early 1950s.

The Lloyd family founded the L&L Service Station at 12900 South State Street. Maurine Lloyd and two of her daughters were photographed in front of the shop in the late 1950s or early 1960s. Unlike the self-service gas stations of today, drivers in the 1950s would pull into a service station and an attendant would fill their gas tanks, clean their windshields, and check their tires and oil as needed.

Leland and Linda Richins owned a Phillips 66 gas station and auto repair shop in Salt Lake City and opened another in Draper in 1965 at the corner of 700 East and 12300 South. They took a creative approach to marketing their new venture and offered whole fresh-baked pies to customers. A family member donned a homemade clown costume to advertise the special giveaway. (Courtesy of Linda Richins.)

In 1940, the State of Utah broke ground for the new state prison near Point of the Mountain. A.R. Mickelsen had advocated for the site with Gov. Henry Blood's advisory board. The farmland that became part of the prison property was originally owned by S.J. Gordon. The first prisoners moved from the Sugar House Prison in Salt Lake City to Draper were housed at first in empty chicken coops. When completed, the Utah State Prison had a capacity of over 4,000 and housed both male and female inmates in separate units. Among the most infamous convicts housed within the prison's walls were serial killer Ted Bundy, later extradited to Colorado, and Gary Gilmore, who was executed on the premises in 1977. Upon its closure and demolition in 2022, only the prison chapel was preserved. (Used by permission, Utah State Historical Society.)

Originally called The Pearl, the Draper Theatre was rebuilt after it burned down in the early 1950s. The Pearl was constructed in 1938 as a movie theater by the Howell family. At the time, serials and newsreels were shown before the main attraction. It was purchased by Charles and Vanessa Nelson and renovated to accommodate live performances in the 1980s. It has been the home of Draper Historic Theatre ever since.

In 1952, Tom Stowe and Dr. J.T. Sorenson built a new drugstore and doctor's office on the northwest corner of 12400 South and 950 East. The drugstore had a soda fountain at first, but it was later removed to make room for an increased inventory of prescription medications.

Draper National Bank and Trust opened its doors in July 1964. It may have been the first bank in Utah to have a drive-up teller window. The bank offered free checking accounts, and employees communicated with customers on a first-name basis. The name was changed to Draper Bank and Trust in 1968. Its basement was rented out for special events and meetings.

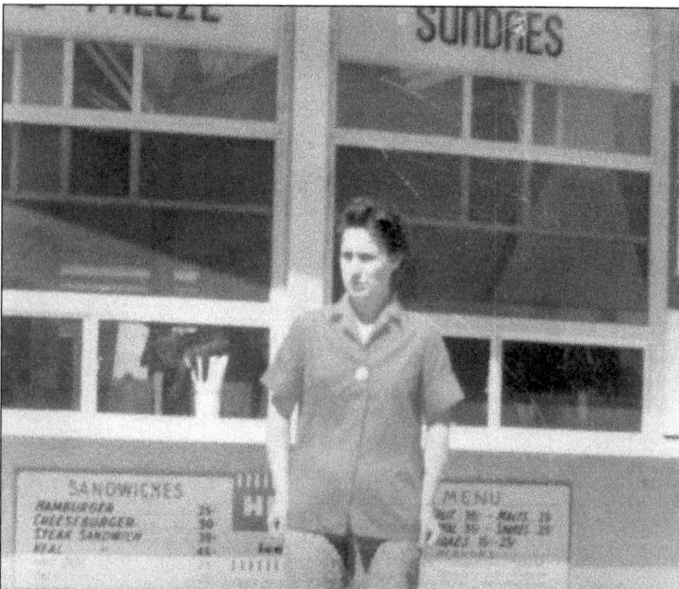

Alice Mae (Sadler) Ballard grew up in Draper and attended Draper Elementary and Jordan High School. She married Allen Ballard in 1940. Having worked her whole life doing everything from candling eggs to waiting tables, Alice Mae was determined to start her own business. The Ballards established Iceberg Drive-In in 1960. Alice Mae worked at the Iceberg for decades before her son Don took over the family business. (Courtesy of Don Ballard.)

The land around 673 East 12300 South, now a busy commercial area, looked very different when this photograph of the home of Allen and Alice Mae Ballard was taken. Seeking to start her own business, Alice Mae met a man named Hap Vitale who had a business model and building design for a fast-food franchise. Ballard's Iceberg Drive-In was built in front of the house in 1960, where it remains today. Over 60 years later, Ballard's Iceberg Drive-In still offers burgers, fries, shakes, malts, sundaes, ice-cream cones, and more to hungry patrons. In the early years, cheeseburgers could be had for 30¢, and a chocolate malt cost only a quarter. (Both, courtesy of Doug Ballard.)

For a brief time in the early 1960s, Draper hosted a roadside attraction, Dinosaur World Sculpture Studio, located at 12601 State Street. Sculptor Elbert Porter had an art studio where he and his team of assistants created life-sized dinosaur statues meant for public education and enjoyment. As they finished each statue, the dinosaurs were put on display outside the studio. This brochure encouraged visitors to see them.

DINOSAURS . . . .

★ FULL SIZE
★ LIFE LIKE

are being sculptured at Dinosaur World Sculpture Studio.

These prehistoric animals range from 6 to 80 feet in length.
They will be placed in a setting depicting their natural surroundings.

You will be amazed at the scientific accuracy of these tremendous animals.

**STOP and see them.**

Salt Lake City

Visit

**DINOSAUR WORLD**

Sculpture Studio

12601 South State Street

After their completion, Porter's dinosaur sculptures were displayed briefly at parks in West Yellowstone, Montana, and Orderville, Utah, before the Division of Utah State Parks purchased them for its museum in Vernal. In 1977, a caravan of trucks brought the statues to their new home on the grounds of the Utah Field House of Natural History State Park Museum, where they remain today. (Photograph by David J. Weinstein.)

# Four

# A HISTORY OF
# DRAPER'S SCHOOLS

An ongoing thread throughout the history of Draper is the value placed on education by the community. In the 1860s, the town was hailed by the leaders of The Church of Jesus Christ of Latter-day Saints as "the cradle of education" in Utah Territory. At various times over the years, Draper schools and teachers have been recognized for their innovative approaches to education as well as a commitment to music and the arts.

In 1852, the settlers of South Willow Creek assembled a school board and built an adobe schoolhouse on part of the north wall of the fort. William R. Terry became the town's first school teacher. There were 25 students in that first classroom, including some of Terry's own children. School took place during the winter months and ended in the spring in time for the planting of crops.

As with most frontier American schools, teachers were not required to have any particular qualifications, training, or background. There was no standardized school curriculum. Students were generally educated in reading, writing, and basic arithmetic.

As the town grew, larger schools were needed. Another adobe building known as the Vestry was constructed within the fort in 1860 and used as both a schoolhouse and church meetinghouse. Two years later, a larger facility known as the Old White Meetinghouse was added to the Vestry, and an extensive bowery was later added on to that. The Vestry was the school where Dr. John Rockey Park first taught in 1861.

John Rockey Park, MD, was educated as a medical doctor at New York University but later decided against practicing medicine as a career and headed west in search of new opportunities. He found work on the farm of Absalom Smith in Draper. When Bishop Stewart learned about Park's level of education, he was immediately hired to teach school.

Park was an excellent and innovative teacher. Word of his talents grew, and soon students came from miles around to attend his classes. Eventually there were over 100 students of all ages and levels of scholastic achievement at the Vestry school. When Brigham Young and other church leaders visited Draper in 1867 and toured the school building, they were amazed by what they saw.

The school was filled with all kinds of exhibits befitting a natural history museum. These included preserved botanical and animal specimens along with insect and butterfly collections displayed on pins. Tables in the school room held samples of stones and minerals organized according to their geological ages and order. There were visual displays showing examples of the work of carpenters, wheelwrights, blacksmiths, shoemakers, gunsmiths, locksmiths, and others as well as gadgets to demonstrate principles of math and physics. Examples of fine penmanship were displayed on the walls. The library was full of books, including rare volumes, and the blackboard covered with programs and notices.

A collection of educational materials such as this could be found nowhere else in the entire territory. Through these materials, Dr. Park went beyond the basics of "reading, 'riting, and 'rithmetic" and opened students' eyes to science and the wider world around them. Dr. Park taught in Draper until 1869 when he left to become the first president of the University of Deseret, now the University of Utah, in Salt Lake City.

As Draper continued to grow, so did the need for more schools. The Central School was built in 1883, followed by the South and East Side Schools in the early 1890s. Still, there were not enough classrooms. At various times, school was held in makeshift classrooms in buildings all over town, including a pool hall and the upstairs portion of the Rideout Store.

Finally, the school board determined to consolidate the various schools into one. The first Park School was built in 1898 on the site of the Central School. Named after Dr. Park, the school brought together eight grades into one building. For the first time, students were transported to the Park School during the winter months in horse-drawn covered wagons called "kid wagons."

The Draper Park School was the first to add a ninth grade, which made it the first rural school in Utah to offer high school–level classes. In 1902, the University of Utah recognized diplomas from the school and allowed the students to attend the university.

By the turn of the 20th century, significant changes had occurred in Utah's school systems. Instead of parents paying for their children to enroll, schools were funded by taxes. Qualifications for teachers became more stringent. By law, the county superintendent conducted annual teachers' examinations to determine pedagogical ability. Based on test scores, teachers received certificates allowing them to teach either grammar school or upper grades.

In 1912, the old Draper Park School was razed, and a new school with 11 classrooms was built in its place. The yellow brick school was built in a Classical Revival style and was designed by the architect Niels Liljenberg. It accommodated students from elementary school through the ninth grade.

One of the first principals of the new Park School, Reid Beck, is said to have had as much impact on education in Draper as Dr. Park. Beck served as principal from 1917 to 1943. Under his guidance and leadership, education in the arts flourished. The music programs at the school, particularly the marching band, became renowned.

In 1927, Beck spearheaded a new program in which students would raise funds to purchase a piece of art of their own choice for the school. The art collection, which later came to be known as the Reid and Willda Beck Collection, has grown over the decades. Noteworthy pieces in the collection include a painting of Ichabod Crane by Norman Rockwell and works by children's book author Eric Carle and the painter and television personality Bob Ross.

Jordan High School, in neighboring Sandy, opened to students in November 1914. Created by the same architect as the Park School, it was designed to draw students from surrounding towns, including Draper. Students rode to Jordan High School in horse-drawn wagons or sleighs depending on the weather and had to leave very early in the morning to get to their 9:00 classes on time. For the first half of the 20th century, Jordan High was the only high school serving the entire southeastern section of Salt Lake County.

Until the 1950s, Draper students attended elementary and junior high school in Draper and commuted to high school at Jordan High. This changed in 1954 when Mount Jordan Middle School was built in Sandy. Draper Park School underwent a massive renovation that year and became solely an elementary school.

By the 1970s, the school was transformed yet again. Draper Elementary was built in 1976 and the old Draper Park School became the site of city hall. To relieve overcrowding at Jordan High School, Brighton High School was constructed in Sandy in 1969, followed by Alta High School in 1978. No high schools were built in Draper until the construction of Juan Diego Catholic High School in 1999, followed by Corner Canyon High School in 2013.

William R. Terry and his family were among the Latter-day Saint pioneers who came to Utah Territory after being driven out of Nauvoo, Illinois, in 1852. Although Terry had only a little over a year of formal education, he was determined that his children should learn to read and write. He became the first schoolteacher in what is now Draper.

Originally built in 1863, the Old White Meetinghouse was used as a schoolhouse as well as a church meetinghouse and community center. Dr. John Rockey Park taught in this building as did some of his students who succeeded him as teachers in Draper. The building was a focal point of the community until 1904 when it was torn down.

John Rockey Park, MD, was born in Tiffin, Ohio, and attended Ohio Wesleyan University. He received his medical degree from New York University in 1857. After a few years of practicing medicine, he had a change of heart about his career and ventured west. He came to Draper where he was hired to teach at the Old White Meetinghouse. At first, Dr. Park's salary as a schoolteacher was comprised of cash, potatoes, and wheat. He provided Draper students with a new curriculum in a wide variety of subjects. In 1869, he became the first president of the University of Deseret, now the University of Utah. Dr. Park's philosophy of teaching, which he set forth in an 1885 speech to future teachers, appears on a plaque at the university: "Always remember in your teaching that the grand purpose of your labors is to make citizens—active, thinking, intelligent, industrious and moral men and women."

A native of Denmark, Peter Niels Garff endured a perilous trek across the prairie to Utah in 1857. Garff and his wife, Antomina Sorenson, transformed 30 acres of uneven land into a productive farm and orchard. Garff served as the Draper Sunday school superintendent for many years and also established the first free public schools in Draper, funding them with tax dollars from the railroad.

Members of the community were photographed outside of the Central School in 1883. Constructed from lumber brought in from Alpine, the school featured cloak rooms, a library that was also used as a museum, a room for storing fuel for heat, and one large room that could be divided into two classrooms. The old Draper Park School now stands on the site.

Both the East Side and South Side Schools were built so that younger children would not have to walk a great distance from their homes on the south side of Draper. Both schools were comprised of one large room with a basement and furnace for heat. The South Side School, shown here, was later turned into a private residence.

When teacher Melvina "Dezzie" Andrus married in 1902, she had to quit her job. At the time, married women were not allowed to teach, ostensibly because pregnancy would prevent them from being able to fulfill their duties. This so-called "marriage bar," which was common throughout the United States, gradually ceased to exist by the 1950s. However, the ban did not officially become illegal until the Civil Rights Act of 1964.

By the fall of 1898, the former Central School was remodeled to consolidate all three of Draper's schools into one. It was renamed Draper Park School in honor of Dr. John Rockey Park. Eight classrooms accommodated eight grades and approximately 30 ninth graders. It is said that its bell was heard all over town.

Graduating students posed with their diplomas from the Park School in 1908. At the time, students were not placed in grades according to their ages or the length of time they had been in school but by their reading level. Classes taught in the ninth grade during this era included bookkeeping, algebra, spelling, Latin, and English.

87

The last class to graduate from the old Park School posed under the front archway in 1912. The fashions worn by the students are typical of the era. The ladies are wearing shirtwaists, button-down blouses modeled after men's shirts that offered greater comfort. The young men are wearing suits and shirts with detachable collars made out of celluloid.

The Draper Park School was built on the same site as the previous school of that name in 1912. Elementary and junior high school classes were held in this building over the years. Listed in the National Register of Historic Places, as of 2015 the school now houses small businesses in its classroom spaces. Coat hooks and shelves that once held lunch pails still line some of the hallways.

ordan High School, built on State Street in Sandy, opened its doors to students in the fall of 1914. The board of education had commissioned architect Niels Liljenberg, also the architect for Draper's Park School, to design the building. Its construction cost was $165,000. Students came to Jordan High School from Sandy, Herriman, Riverton, West Jordan, Midvale, Granite, Draper, and other smaller towns. This view of the parking lot shows early automobiles along with the horse-drawn "kid wagons" that transported students from nearby towns. In the cold weather months, Draper students often brought hot bricks with them to keep their feet warm on the hour-long ride to school. Students who rode their own horses to school had to furnish grain for the animals during the school day. The school stood for over 80 years before being demolished in 1996. A new, larger high school was built nearby, and the Jordan Commons movie theater and restaurant complex opened on the site of the original school in 1999.

In 1917, J.R. Allen, who served on the Jordan District School Board, was so impressed with a Provo teacher and principal named Reid Beck that he offered him the position of principal at the Draper Park School. Beck hesitated at first but decided to make the move to Draper after Allen offered him a raise in salary and the privilege of bringing eight teachers to the district with him. Beck was originally from Spring City in Sanpete County and studied to become a teacher at Brigham Young University. Beck instilled a sense of mutual cooperation between himself, the faculty, and the students at the Draper Park School. He stayed on as principal until his untimely death in an automobile accident in 1943. In this photograph, Beck, wearing a tie and vest, is seated in the first row with the 1917-1920 faculty of the Draper Park School around 1920.

Built by the Green family around 1900, this Victorian-style home was later owned by the Jordan School District to provide housing for educators. It was the home of Reid Beck during his tenure as principal of the Draper Park School. The home is listed in the National Register of Historic Places and is currently a private residence. (Photograph by LaRayne Day; courtesy of Draper Historic Preservation Commission.)

Reid Beck posed with a group of some the youngest students at the Draper Park School and their teacher on April 23, 1919. Teaching methods for primary grades were somewhat different from today in that more emphasis was placed on memorization. The students pictured here would have been expected to learn and recite short poems and songs in the classroom.

Classrooms of the 1920s looked much different from those of today. This room in the Draper Park School featured rows of desks that were often equipped with inkwells. Students wrote in their notebooks with dip pens, although fountain pens became more widely used by the 1920s. The walls were lined with blackboards. Students would be asked to clean the boards and erasers from the chalk dust.

The Great Depression meant that the tax revenue that kept schools afloat dried up. The Draper Park School was closed early for lack of funds in 1932. Clayton T. Vawdrey, a high school senior in that year, recalled that some students went to night school and summer school to earn their diplomas. In the face of hardship, going to school and earning a degree was the path to success.

The Draper Park School was renovated and enlarged in 1928. The addition provided more classrooms, an auditorium, workshops, and spaces for domestic science classes, a music room, stage, locker space, showers, and restrooms. The new additions highlight evolving attitudes about education, which gave students the opportunity to study the arts in addition to core academic subjects.

A principal who knew the names of all of the students at the Draper Park School, Reid Beck (center, first row) has been compared to Dr. Park for the lasting impact he made on education in Draper. During his tenure as principal, the school's band program achieved acclaim, and the students started an art collection that now bears his name. Beck is pictured here with some of the Draper Park School teachers during the 1930s.

In 1938, the Works Progress Administration and Jordan School District cosponsored the painting of a large mural over a doorway at the Draper Park School. Artist Paul Smith sought to capture the history of education in Draper in the work, depicting the Draper of pioneer days on the left and prominent citizens of the town in the 1930s on the right. The central figure in the painting is Dr. John R. Park surrounded by William R. Terry, Bishop Stewart, and other early settlers. The old adobe fort is seen on the left along with livestock, a plow, the Old White Meetinghouse, and the John R. Park Building at the University of Utah. Draper industries of the 1930s and the Draper Park School are portrayed on the right side of the mural along with Principal Beck and school board members Soren Mickelsen and J. R. Allen. In 2014, the Draper Visual Arts Foundation arranged for the preservation of the mural and moved it to Corner Canyon High School.

In the 1930s, many students at the Draper Park School participated in school plays in their new auditorium. Judging from the windmill in the background and the Dutch caps worn by the students, this production from 1938 was likely *Hans Brinker, or the Silver Skates*, which was a popular story at the time. The arts were an important part of school during Principal Reid Beck's tenure. (Courtesy of Shelley Feuerstein.)

The Draper Junior High School band was photographed in front of the school in 1931. The band grew in scope and talent during the 1930s thanks to inspiring band teachers ElRay L. Christiansen and Murray R. Lewis. They performed at the dedication of the Orrin Porter Rockwell Monument at the Point of the Mountain in 1932.

The Draper Junior High School marching band practiced on the lawn in front of the Draper Park School in the mid-1930s. The band performed at civic events and parades, including the big Fourth of July celebration in Provo. The young musicians proudly wore new black and white band uniforms, which were designed and sewn by members of the Draper community. The uniforms were black slacks with white blouses and military-style caps. Members of the marching band worked hard and had to memorize their music in addition to learning to march in step. When ElRay Christiansen moved to Logan in 1936, Murray R. Lewis replaced him as the band teacher. Lewis was a no-nonsense teacher who encouraged his students with the phrase, "Practice doesn't make perfect! The right kind of practice makes perfect!"

The World War II and postwar era of American education was distinguished by efforts to increase literacy and set nationwide standards for education. In 1942, a Miss Parker's second graders at the Draper Park School studied math, science, social studies, handwriting, and reading. The Dick and Jane series of readers was popular during this era. Elementary students participated in spelling bees and math contests.

From left to right in this photograph from the *Deseret News*, Dallas Dean, Donna James, Jean Jensen, Shirley Nelson, and Rachael Zitting printed out copies of the Jordan High School newspaper, *The Broadcaster*, on a mimeograph machine during the 1946–1947 school year. Jean Jensen later graduated from BYU and married Robert Hendricksen. She taught elementary school in the Jordan School District for over 32 years and cofounded the Draper Visual Arts Foundation. (Courtesy of Jean Hendricksen.)

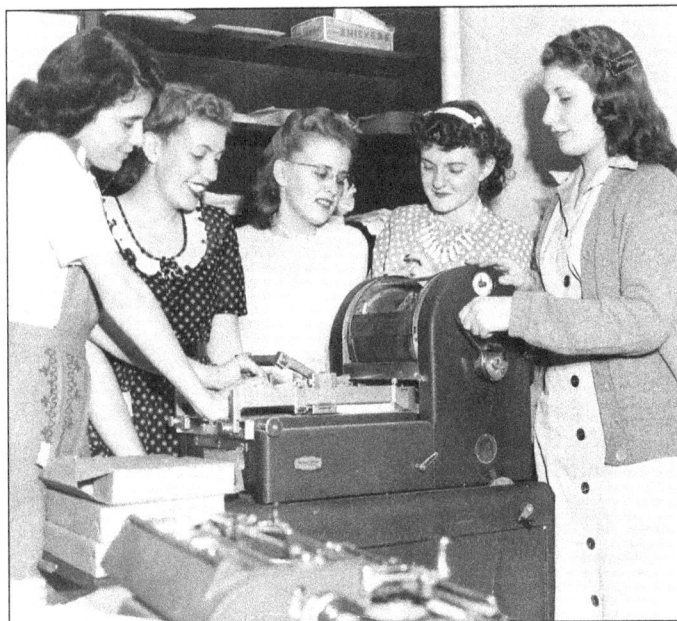

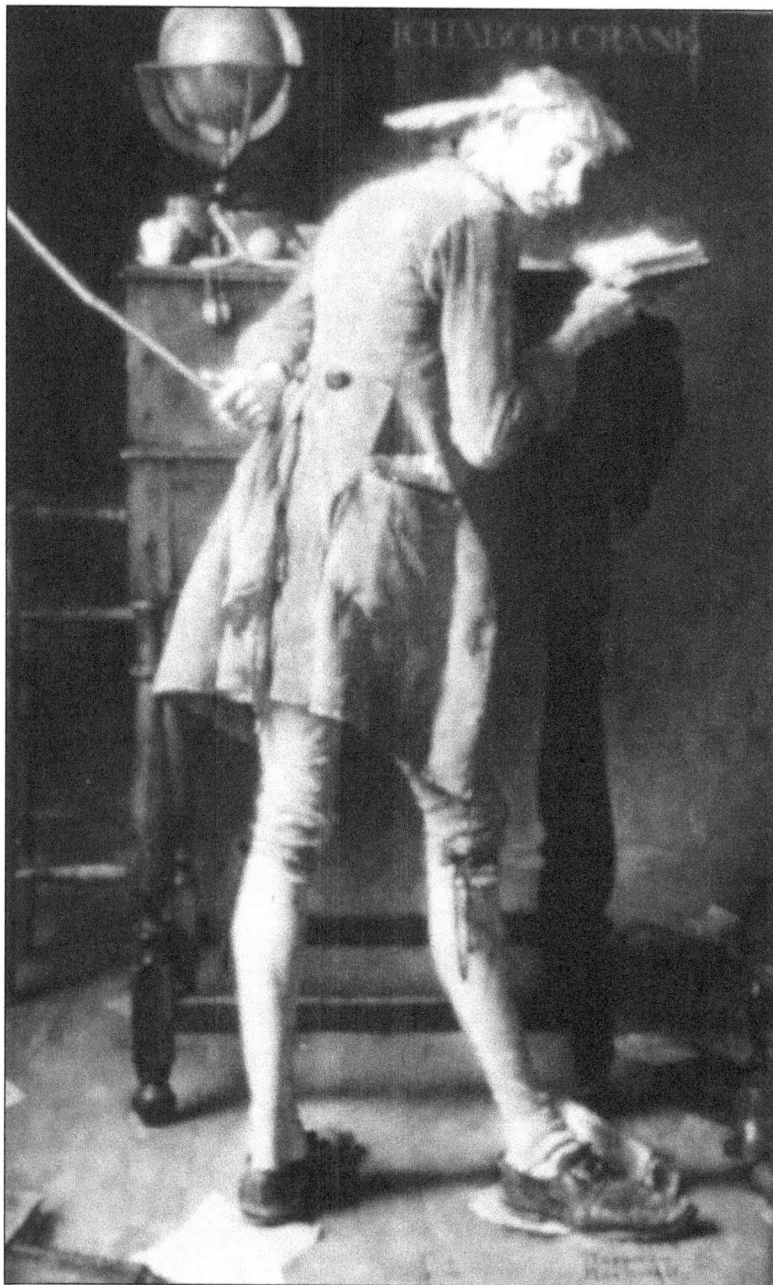

Principal Beck started the tradition of Draper students traveling to the Springville Museum of Art to purchase a work of art for the school in the 1920s. In 1951, the graduating class at Draper Junior High School selected Norman Rockwell's painting *Ichabod Crane* as their contribution to the collection. The painting of Washington Irving's ill-fated schoolteacher was out of reach financially until Rockwell learned that schoolchildren wished to purchase it and lowered his price. With the encouragement of their class leaders, Beatrice Hill, Luana Lunnen, Janet Duffin, Shirlene Day, and LaMar Walbeck, the class held fundraisers and worked odd jobs to raise money to buy the painting. It is one of the outstanding pieces in the Reid and Willda Beck Collection. (Courtesy of Canyons School District and Draper Visual Arts Foundation.)

DRAPER SCHOOL

Mrs. Beck

Grade 1

1954-1955

Reid Beck's widow, Willda Beck, taught elementary grades for many years at the Draper Park School. Her first graders in 1954 enjoyed new renovations to the school, including updated classrooms and a state-of-the-art cafeteria. While in past decades students brought their lunches from home, Draper Park School students in the 1950s could have hot lunches in the cafeteria.

At graduation, the Jordan High School class of 1961 celebrated the 50th anniversary of the school's move to its first building. Having started out in a church basement in Midvale in 1907, Jordan High was known at first as the Peoples' College because it accepted students of all ages. By 1961, a total of 1,440 students attended Jordan High School. (Courtesy of Alona Holm.)

The elementary school children (above) photographed in front of the Draper Park School in 1976 were standing on the cusp of great change in their community. Draper officially became a city the following year in 1977. The old school, which had seen generations of Draper children fill its hallways and classrooms since 1912, became the new city hall. Draper Elementary School, seen below in a photograph from the late 1970s, opened in the middle of the school year in 1976. On moving day, the students formed a parade and carried their tote trays up the street from the old Park School.

# *Five*

# CHURCH, PASTIMES, AND CELEBRATIONS

Bound together by a shared faith and the need to survive, the Draper of pioneer days was very close-knit. This communal spirit remained throughout the town's history. The people of Draper worshiped, worked, and played together. Even in times of hardship, they made time for music, dances, sports, and theatrical entertainment. As a community, the people of Draper also cultivated a sense of civic pride and pageantry, as seen in the parades and celebrations commemorating their heritage. The extent to which The Church of Jesus Christ of Latter-day Saints was the cornerstone of the Draper community cannot be understated. Beth (Wight) Fairbourn described growing up in Draper in the 1930s, "When we grew up, the church and the town was one and the same. It was our life, outside of the home."

The lives of Draper's Latter-day Saints were structured not only around their families, farms, and livelihoods but sacrament meetings and church organizations such as Sunday school and the Relief Society. Young men in the community attended priesthood classes, and young women went to meetings of the Mutual Improvement Association. Younger children received religious education at Primary. Many social activities such as dances, outings, picnics, and sports were also part of Church life.

To be sure, there were residents of Draper over the course of its history who were of different faiths, notably Lutherans and Roman Catholics. However, members of these religions had to travel outside of Draper to attend worship services.

Music was an essential part of life in Draper. From the very beginning, the pioneer settlers made singing part of their worship and held dances. Generations of the Boulter and Orgill families in particular produced bandleaders whose musical groups performed at dances and civic events throughout the South Valley well into the 20th century.

Written histories of Draper are filled with anecdotes about dances. In pioneer times, rugs were rolled back on cabin floors, and the settlers danced until dawn. Dances were held outdoors in moonlit orchards, at Rideout Hall, and later in the First Ward roundhouse. The people of Draper worked hard but also knew how to have a good time.

Rideout Hall was also the home of theatrical performances. Touring theater companies came through Draper along with other bygone forms of entertainment. "Moving panoramas," in which stories were told using an illuminated painting rolled out to the accompaniment of music and narration, were popular before the era of motion pictures. Draper's first community Dramatic Club was established there in the 1880s.

In 1876, the first team sport organized in Draper was baseball. While baseball was predominantly a game for young men, both girls and boys played basketball in Draper in the early 1900s. The Boy Scouts and later The Church of Jesus Christ of Latter-day Saints put together basketball leagues starting in the 1920s. Wards held basketball and baseball competitions throughout the 20th century. The 1930s saw the rise of school football teams as well as softball and baseball teams sponsored by local businesses. Community-sponsored Little League baseball began in the 1950s.

Winter sports such as sledding and ice skating have been enjoyed in Draper since the 19th century. Skiing first began to catch on in the 1920s when the woodworking teacher at the junior high school taught students how to build their own skis.

Scouting was a big part of Utah culture in the 20th century. In 1913, a 105-year-long partnership between The Church of Jesus-Christ of Latter-day Saints and the Boy Scouts of America was established out of a shared commitment to teaching morals and responsibility to boys. Draper's first Boy Scout troop was founded that same year. The Scouts worked on service projects and went hiking and camping.

In the 1930s, Draper was the only community in the area to have an adult Scouting group for men and women, which included most of the town's leading citizens. The members helped younger Scouts earn merit badges, sponsored community events, learned first aid, and conducted their own service projects.

Social clubs and service organizations became increasingly popular in Draper in the decades following World War II. The American Legion, Lions Club International, and later the Lady Lions Club attracted many members. Many young people became active in 4-H Club in the 1950s and 1960s. For a small agricultural town, there were always things to do in Draper.

Looking back over the history of the town, it is clear that the people of Draper took great pride in their community and enjoyed the pageantry of civic celebrations. The first parade in Draper was organized by Dr. Park in the 1860s to welcome Brigham Young to the town. The church leader was met by a group of flag-bearing men on horseback and a brass band with 100 schoolchildren and 30 boys carrying wooden guns marching along in unison. It was a remarkable display that made an impression on Young.

By far the most popular community event in Draper was Pioneer Day. Just two years after Brigham Young and the first settlers arrived in the Salt Lake Valley on July 24, 1847, the date was formally observed and celebrated. In 1857, Draper townspeople were among thousands who gathered to observe the 10-year anniversary at Silver Lake in Big Cottonwood Canyon. Pioneer Day or "the 24th," as it was called in Draper, has been celebrated in Utah ever since with parades, rodeos, games, and other special events. It is a patriotic holiday recognizing the freedoms enjoyed by Utahns and commemorating the achievements of their forebears.

The first Pioneer Day celebrations held at Draper Park began in 1902. Reenacting pioneer times was the theme with people wearing period costumes and riding pioneer wagons in the parade. One year, George Terry, whose mother was Native American, helped stage a reenactment of a battle between "settlers" and "Indians," complete with a gunfight and the burning of a wagon.

Over the course of the 20th century, "the 24th" celebrations in Draper included the selection of a Draper queen, a parade, rodeo, horse races, ball games, contests, fireworks, and plenty of refreshments. A drawing to win an automobile was a popular tradition for many years. As the rural community transformed into a city and a new century began, "the 24th" evolved into a weeklong celebration in July known as Draper Days and Draper Nights.

In the 21st century, the Draper Days parade is among the largest and well-attended parades in the state. Despite the dramatic changes and growth that have occurred in Draper over the decades, the community's pride in its history is very much alive.

Brigham Young founded the Young Ladies' Department of the Ladies' Cooperative Retrenchment Association as a means to educate and encourage spiritual growth in young women in 1870. Seven years later, the organization was renamed the Young Ladies' Mutual Improvement Association. The members of the first Draper chapter, seen in this photograph from the 1880s, met at the home of Zebulon and Julia (Fitzgerald) Stewart. At the time, the Draper chapter had no manuals, so the meetings consisted of testimonies, readings, and essays on a variety of subjects. In 1915, the Church established its first recognition program for young women in the "Beehive" handbook, which required participants to accomplish tasks of personal improvement. In the 21st century, Young Women is a Church organization dedicated to helping teen girls to grow spiritually, socially, physically, and intellectually.

Draper's Sunday school was among the first to be organized in Utah Territory in the spring of 1857. Open to Latter-day Saints of all ages, its goal was to strengthen faith through the teaching and learning of doctrine and fellowship. Social outings and picnics in the outdoors have been a part of Draper's Sunday school from the beginning, as seen in this photograph from 1884.

Generations of Draper's Latter-day Saints were baptized in this pond behind the Lauritz Heber Smith house. It remained a popular place for baptisms well after the first ward chapel was constructed. Baptisms were held after the corn was high enough to allow the boys to change out of their wet clothes amid the cornstalks. Girls changed in the barn. (Courtesy of Esther Kinder.)

Roy Boulter shared his pioneer father's passion for music. He played the bass fiddle in a band that performed at dances and gatherings throughout the South Valley. From left to right, Boulter, A.R. Ballard, William B. Norris, Erastus Rasmussen, Woodruff Sorenson, an unidentified musician, Stanley Rasmussen, and John Boberg were photographed with their instruments in the 1890s.

Carl William Hendricksen was born in Draper in 1866. A carpenter by trade, he played the banjo, mandolin, and steel and Spanish guitars. Hendricksen was well-known for his talents and played in the Boulter and Orgill bands. When his first wife, Anna, passed away, Hendricksen married Jessie (Shaw) Bradley, an Englishwoman who played the violin.

Draper's first community theater group, which performed at Rideout Hall, began in the 1880s and lasted for many years. The Dramatic Club presented a variety of plays, including popular melodramas of the era such as *The Octoroon* and the temperance drama *Ten Nights in a Bar-Room*. The stage was set with stock scenery painted by a local artist.

A pioneer who loved dances and hosting parties, Willard Snow initiated the first Old Folks Day celebration, which took place in 1898. It was a dinner with speeches and entertainment to honor the original settlers of the community. An annual event in Draper for over 90 years, Old Folks Day later evolved into the Heritage Banquet hosted by the Draper Historical Society.

The history of Draper's queens goes back well over a century. This photograph from 1900 shows the queen in a royal robe with her attendants. Her name is unidentified, but a note written on the back of the photograph suggests she was a member of the Rasmussen family. As of the 2020s, the competition has evolved from being a beauty pageant to a scholarship program.

Lavona Rasmussen smiles from her seat at the piano in her parents' parlor in this photograph from the early 20th century. Her father, Soren Rasmussen, managed the Draper M&M store. A lifelong musician, Lavona grew up to become the church organist in her stake, a position she held for 65 years.

On May 24, 1903, Draper's First Ward Chapel was dedicated by John R. Winder, the first counselor in the First Presidency. Built by Draper residents over a period of 10 years at a cost of $13,000, the chapel was located in the area where the old fort had been. The chapel was constructed out of red bricks and granite cut from the mountains. The interior featured large windows with stained-glass tops, a baptismal font, comfortable seating, and classroom space in the basement. Large concrete steps led to the front entrance. Inside the front door was a long vestibule where latecomers to church would wait before entering the chapel. The church yard was fenced in with gates on the north and south sides. The Draper Round House, a recreation center, was added in 1916. The First Ward Chapel stood for over 50 years.

Teachers and officers of Draper's Sunday school posed for a photograph outside the new First Ward Chapel in 1910. In that year, Soren Rasmussen, first row, second from the left, became the Draper Ward bishop. A culinary water system, electricity, telephones, and automobiles came to Draper in the 1910s, but World War I loomed over the decade. Forty-eight young men from Draper fought in the war, and not all of them returned home.

Both girls and boys played basketball in Draper in the early 1900s. Their uniforms were quite different from the shorts and jerseys worn today. The boys wore button-down shirts and long padded trousers that extended below the knee while the girls wore wool serge suits that also covered the knees. The Draper basketball team won a trophy in 1908.

The "kid wagons" that transported students to school were decorated as parade floats for the Fourth of July and 24th holiday festivities. In this photograph, students wore their Sunday best to ride this star-spangled patriotic wagon. The picture was taken after 1912 as the flag has 48 stars, showing the inclusion of the new states of New Mexico and Arizona.

Draper Park was officially incorporated in 1901 after Chris Hendricksen and D.O. Rideout purchased the land from Joshua Terry. A grandstand was built on the property shortly thereafter. Horseshoe pitching, baseball games, and buggy racing, as seen in this undated photograph from the early 20th century, were among the activities that took place there.

Saltair, an amusement complex constructed on the Great Salt Lake in 1893, was intended as a western counterpart to Coney Island. Many families in Draper, especially young people, took the train to Saltair on summer days. From left to right in this photograph from 1913, (seated) Leona Allen, Inez Stringfellow, (standing) Eva Allen, and Louise Allen posed in their fashionable hats for a trip to Saltair. (Courtesy of Paul Smith.)

Team sports in Draper began in 1876, the same year that the National League of Professional Baseball was established in New York City. Baseball fever spread across America, all the way to Utah Territory. Draper's first baseball team called themselves the Red Sox. By 1915, when this photograph was taken, Draper's baseball program went by the name of the Draper Ball Club.

DRAPER

Utah became a state in 1896. In the early 20th century, the Fourth of July was celebrated in Draper with a public reading of the Declaration of Independence, followed by picnics, games, fireworks, and sometimes a parade. This photograph, which has the date July 4, 1929, written along the bottom, shows horse-drawn parade floats going south on Fort Street.

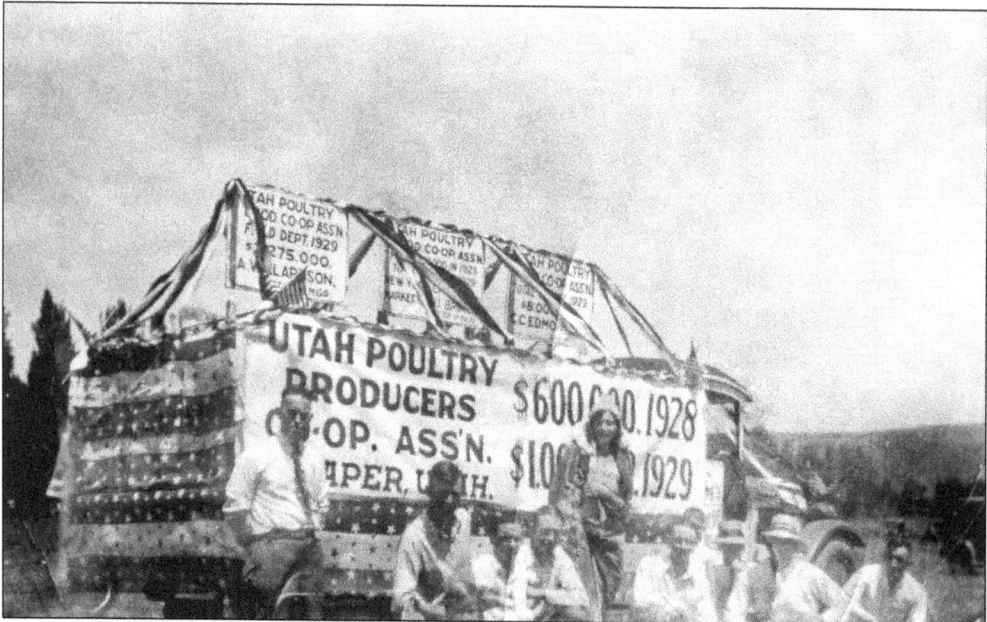

The floats that comprised the Pioneer Day parades in Draper celebrated not only the town's pioneer past but were a form of advertising as well. The egg and poultry business in Draper was growing rapidly in the summer of 1929. The signs on the parade float in this photograph proclaim the financial success of the burgeoning industry.

The streetcar cables and buildings seen in the background of this 1931 photograph of a Draper pioneer-themed parade wagon indicates that it was not taken in Draper but at a Pioneer Day parade in Salt Lake City. As of 1931, the celebrations surrounding July 24 in the state capital were known as Covered Wagon Days. Church wards, stakes, and community groups from all over the valley participated.

On June 13, 1935, members of the Draper community donned their pioneer costumes for a celebration commemorating the Pony Express. Prior to the invention of the telegraph in 1861 and later the railroad, which rendered it obsolete, Pony Express riders carried mail from points East to the West Coast. In pioneer days, many Utahns worked for the Pony Express and a station was located at the Point of the Mountain.

The Friendship Club, as seen in this photograph from the 1930s, was a social club for mostly senior ladies. The club was most active during the mid-20th century. Friendship Club members met regularly for social gatherings, reached out to those who were ill, and remembered each other's birthdays with greeting cards.

The Relief Society is a philanthropic and educational women's organization of The Church of Jesus Christ of Latter-day Saints. Members of the Relief Society aid church members in need by nursing the sick, supplying meals, cleaning homes, and providing other help as needed. These members of Draper's Relief Society were photographed wearing their pioneer costumes for "the 24th" sometime in the 1930s.

In the Depression years, young people in Draper filled their days with school work and chores at home and on the farm. For fun, children and teens played sports and games, swam in the irrigation canals, and improvised their own activities. One year, the children who lived on Relation Street decided to celebrate Father's Day with a parade. They made banners and costumes, and everyone who lived on the street turned out to watch the procession. Relation Street was named for the fact that the families who lived on it at the time were related to each other. (Courtesy of Anne (Garfield) Covington.)

When the population of Draper's first ward grew to 1,200, the Mount Jordan stake presidency determined that a second ward was needed. The Draper Ward was divided in 1935, and Heber J. Smith was ordained bishop of the Draper Second Ward. Construction on the new chapel, located at the corner of 12900 South and 1300 East, began in 1936 on land given by Chris and Edythe Hendricksen. W. Cyrus Vawdrey served as the contractor and had five full-time employees working on the project. The Church paid for 70 percent of the building cost and members of the ward paid the balance. Many in the community donated their labor as well. During the years that the chapel was under construction, members of the Draper Second Ward met at the Draper Park School. After years of hard work and sacrifice, the Draper Second Ward Chapel was dedicated in 1941. (Courtesy of Alona Holm.)

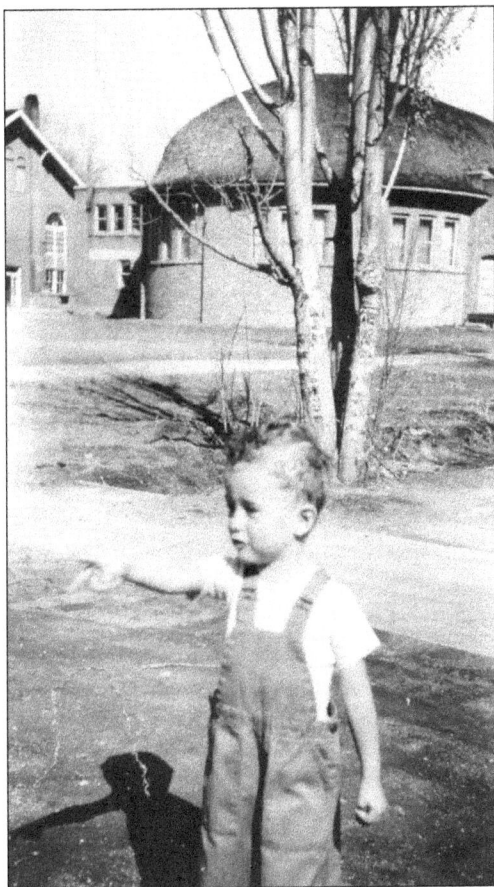

Young Draper resident Doug Ballard was photographed in front of the Draper Round House in the 1940s. The Draper Round House was built in 1916 by members of the community and replaced Rideout Hall as a space for cultural events and recreational activities. It was designed to look like a smaller version of the Salt Lake Tabernacle and became the cultural center of Draper for many years. Wedding receptions, ward dinners, missionary farewells, basketball games, movie nights, roller skating parties, and stake dances took place on its spacious hardwood floor. The photograph below of the Draper Round House looking empty may have been taken shortly before it was demolished in 1961. The Draper Historic Park gazebo was later constructed on the site. (Right, courtesy of Doug Ballard; below, courtesy of Draper Historical Society.)

Founded in Salt Lake City in 1901, the Daughters of Utah Pioneers (DUP) organization was based on other national lineage societies such as Daughters of the American Revolution. Draper's chapter of the DUP celebrated the centennial of the arrival of the Latter-day Saints in Utah with this parade float in 1947. The float was photographed in front of Rasmussen's store.

Many men and women in Draper have served their country bravely in times of war. Veterans and members of the American Legion—(from left to right) Howard Ballard, Owen Nelson, Verner Brynolf, and Bert Nichols—proudly carried the flag in a July 24 parade in the 1950s. A post of the American Legion was established in Draper in 1946.

Draper's Boy Scout Color Guard proudly marches in the "24th" parade in this photograph from the 1950s. The Boy Scout Color Guard hold and post the US flag and the troop flag during flag ceremonies and lead the Pledge of Allegiance at the opening of special events. The Boy Scouts of America have a long history in Draper.

This photograph from the *Salt Lake Tribune* shows a baby contest held in Draper in 1956, possibly sponsored by the Lady Lions Club, which began holding them in the 1940s. The cutest baby with the most cheerful temperament won a silver cup. As reflected in this picture, the Draper community gradually became more diverse after World War II. (Used by permission, Utah State Historical Society.)

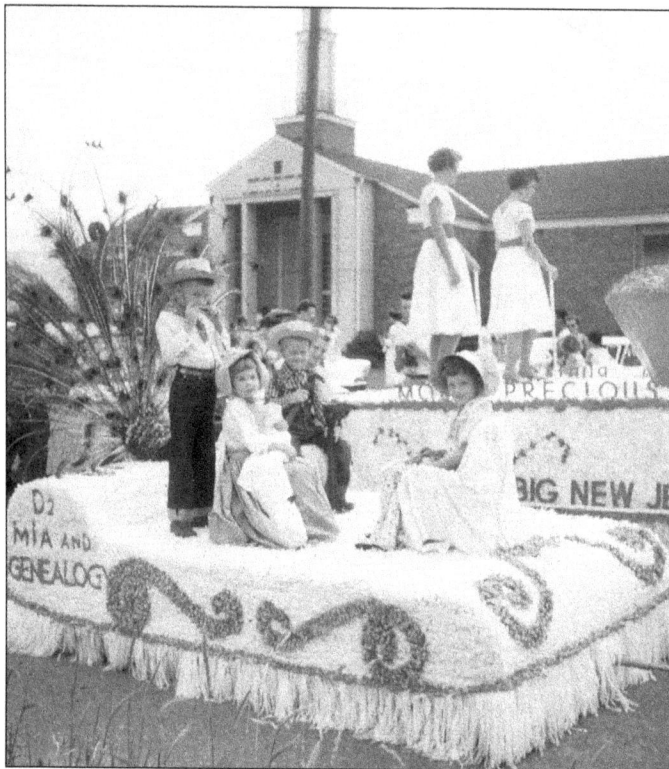

These children were photographed on a Pioneer Day parade float parked in front of the Draper Second Ward in 1959. They were dressed as pioneers to depict the settlers who founded the community. The study of genealogy and learning family history is very important to members of The Church of Jesus Christ of Latter-day Saints. (Courtesy of Alona Holm.)

"It's Smart to Take Part" reads the banner on this parade float from 1959 illustrating some of the activities for young Latter-day Saints of the era. A girl wears her Mutual Improvement Association "bandelo" decorated with emblems earned for different achievements, while the basketball player behind her represents his ward team. Members of the Scouts and children from Primary are also seated on the float. (Courtesy of Alona Holm.)

Ice skating has been a favorite winter pastime in Draper for over 100 years. The town's irrigation canals and many ponds provided plenty of good locations for skating. In this photograph from 1961, Charles and Kathy Smith enjoyed ice skating on the pond at the Terry farm. (Courtesy of Helene Terry.)

Little League Baseball was officially incorporated as a nonprofit organization in Pennsylvania in 1950. Draper's Little League program began just a few years later. Ray Terry started coaching for Draper's Little League program in the early 1960s when this photograph was taken. A plaque near the ball fields in Draper Park commemorates the individuals who established community baseball. (Courtesy of Helene Terry.)

Santa Rides the Fire Truck, an annual event in which one of Draper's volunteer firefighters would dress up as Santa Claus and distribute bags of candy and peanuts to children all over town, was the brainchild of Fire Chief James Rayburn Dow in the early 1950s. The firefighters pitched in to purchase a secondhand Santa suit, and the Salt Lake County fire chief authorized the use of the fire truck. To fund the event, Dow organized turkey shoots that were held twice a year at Thanksgiving and New Years on the sand hill. The tradition lasted until Draper's population became too large. Dow was photographed in front of the fire truck while his granddaughters visited with Santa in 1971. In addition to Santa Rides the Fire Truck, Dow also managed Draper's fireworks shows for the July holidays for many years.

The "24th" parade float for the Draper queen of 1967 reflects the optimism of the Atomic Age when it was widely believed that atomic energy would revolutionize the world. Draper queen Pauline Smith (standing) was photographed riding the float with her attendants Ramona Vawdrey (front) and Kayleen Fitzgerald (back). (Photograph by Lyn Kimball.)

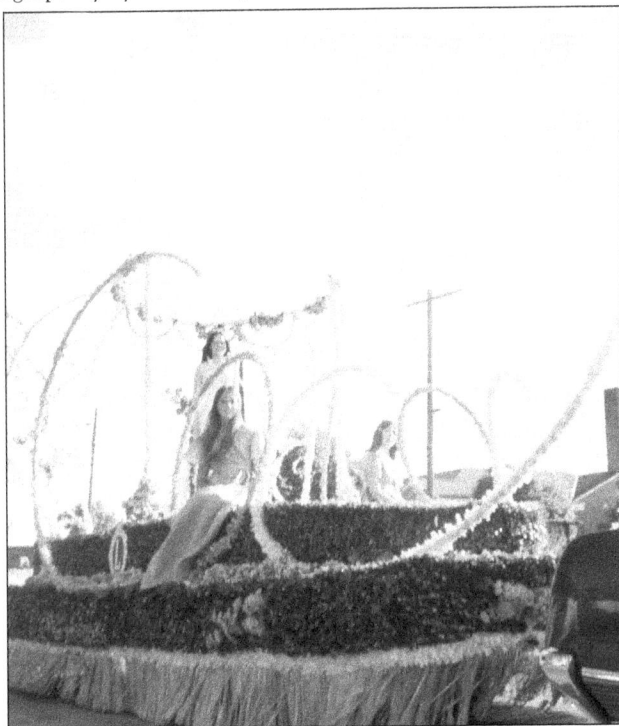

In contrast to the young women depicted in the previous photograph, the long hairstyles and flowing gowns worn by the Draper queen of 1972, Carol Garfield (standing), and her attendants, Janette Landeen and Mary Alice Ballard (back), reflect dramatic changes in women's fashion thanks to the counterculture movement of the era. (Courtesy of Alona Holm.)

**6th ANNUAL**
# WIDOWMAKER
SPORTSMAN
# HILLCLIMBS

## Sun. Mar. 23 & Mar. 30
### WIDOWMAKER HILL
**NORTH EAST OF POINT-OF-THE-MOUNTAIN**
25¢ **CLIMB STARTS 1:00 P.M.**
Sponsored by the
Bees Motorcycle Club

This program cover is a piece of memorabilia from Draper's famous "Widowmaker" hill climb, which was held each spring from 1963 to 1988. Sponsored by the Bees Motorcycle Club, the Widowmaker involved motorcyclists speeding to the top of a steep hill near Point of the Mountain as fast as possible. In 1972, when the photograph below was taken, approximately 20,000 people attended. The photograph, taken from the top of the hill, shows the steepness of the course and the vast numbers of spectators with their vehicles below. Draper's Widowmaker was one of the most challenging tracks in the world and was featured on ABC's *Wide World of Sports*. As the city of Draper continued to become more developed, concerns were raised about the crowds and the environmental impact of the hill climb, which led to its demise. (Both, courtesy of Bill Allinson.)

From left to right, Draper Second Ward bishops Heber J. Smith, Otis A. Pierce, George B. Roden, Arnold G. Adamson, Kay L. Smith, J. Nathan Smith, and Bruce D. Washburn were photographed behind the chapel pulpit in 1976. The relief sculpture they are standing in front of, *Christ at the Well*, was created by the Norwegian Latter-day Saint sculptor Torlief Knaphus. Knaphus also sculpted the *Handcart Pioneers Monument* in Salt Lake City.

To commemorate the nation's bicentennial, an old-fashioned steam engine made its way through Draper on the Fourth of July to the delight of interested onlookers. However, when cinders from the engine landed in the grass next to the tracks, Draper's volunteer fire department had to follow behind putting out the fires. (Photograph by Lyn Kimball.)

Point of the Mountain became known worldwide as one the best places for hang gliding when the sport became popular in the 1970s. To this day, paragliders come to Point of the Mountain for its consistent winds and the gentle grade on its southern slope. In this photograph, Draper resident Mike Westbrook took to the sky in 1976. (Courtesy of Holly Westbrook.)

Community Halloween parties were a Draper tradition since the days of Rideout Hall. These children dressed up for the Halloween party hosted by the Draper Civic Club in 1981. At the time, the annual party was held in the basement of the Park School/city hall as a safe alternative to trick-or-treating. Families enjoyed a costume contest, movie, and a "spook-alley" in addition to plenty of treats.

# BIBLIOGRAPHY

Bradley, Martha Sonntag. *Sandy City: The First 100 Years*. Sandy City Corporation, 1993.
Draper Historical Society. *People of Draper 1849–1924, History of Draper, Utah, Volume One*. Second Edition. Salt Lake City: Agreka Books, 2006.
———. *Sivogah to Draper City 1849–1977, History of Draper, Utah, Volume Two*. Salt Lake City: Agreka Books, 2001.
———. *People of Draper 1849–1932, History of Draper, Utah, Volume Three*. Salt Lake City: Agreka Books, 2004.
*Utah History Encyclopedia*. Salt Lake City: University of Utah Press, 1994.

# DISCOVER THOUSANDS OF LOCAL HISTORY BOOKS FEATURING MILLIONS OF VINTAGE IMAGES

Arcadia Publishing, the leading local history publisher in the United States, is committed to making history accessible and meaningful through publishing books that celebrate and preserve the heritage of America's people and places.

Find more books like this at
**www.arcadiapublishing.com**

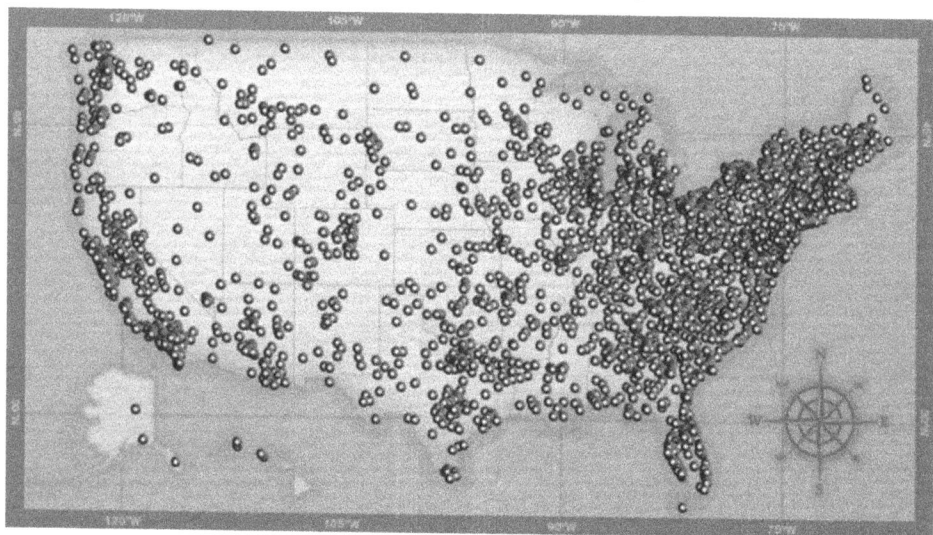

Search for your hometown history, your old stomping grounds, and even your favorite sports team.

www.ingramcontent.com/pod-product-compliance
Lightning Source LLC
Chambersburg PA
CBHW070412100426
42812CB00005B/1717